Key Challenges in Geography

EUROGEO Book Series

Series Editors

This book series addresses relevant topics in the wide field of geography, which connects the physical, human and technological sciences to enhance teaching, research, and decision making. Geography provides answers to how aspects of these sciences are interconnected and are forming spatial patterns and processes that have impact on global, regional and local issues and thus affect present and future generations. Moreover, by dealing with places, people and cultures, Geography explores international issues ranging from physical, urban and rural environments and their evolution, to climate, pollution, development and political economy.

Key Challenges in Geography is an initiative of the European Association of Geographers (EUROGEO), an organization dealing with examining geographical issues from a European perspective, representing European Geographers working in different professional activities and at all levels of education. EUROGEO's goal and the core part of its statutory activities is to make European Geography a worldwide reference and standard. The book series serves as a platform for members of EUROGEO as well as affiliated National Geographical Associations in Europe, but is equally open to contributions from non-members.

The book series addresses topics of contemporary relevance in the wide field of geography. It has a global scope and includes contributions from a wide range of theoretical and applied geographical disciplines.

Key Challenges in Geography aims to:

- present collections of chapters on topics that reflect the significance of Geography as a discipline;
- provide disciplinary and interdisciplinary titles related to geographical, environmental, cultural, economic, political, urban and technological research with a European dimension, but not exclusive;
- deliver thought-provoking contributions related to cross-disciplinary approaches and interconnected works that explore the complex interactions among geography, technology, politics, environment and human conditions;
- publish volumes tackling urgent topics to geographers and policy makers alike;
- publish comprehensive monographs, edited volumes and textbooks refereed by European and worldwide experts specialized in the subjects and themes of the books;
- provide a forum for geographers worldwide to communicate on all aspects of research and applications of geography, with a European dimension, but not exclusive.

All books/chapters will undergo a blind review process with a minimum of two reviewers.

An author/editor questionnaire, instructions for authors and a book proposal form can be obtained by contacting the Publisher.

More information about this series at https://link.springer.com/bookseries/15694

Theresa Bourke · Reece Mills · Rod Lane
Editors

Assessment in Geographical Education: An International Perspective

eurogeo

Springer

Editors
Theresa Bourke
Faculty of Creative Industries, Education and Social Justice
School of Teacher Education and Leadership
Queensland University of Technology
Brisbane, QLD, Australia

Reece Mills
Faculty of Creative Industries, Education and Social Justice
School of Teacher Education and Leadership
Queensland University of Technology
Brisbane, QLD, Australia

Rod Lane
Macquarie University
Sydney, NSW, Australia

New Zealand School Trustees Association
Wellington, NSW, Australia

ISSN 2522-8420 ISSN 2522-8439 (electronic)
Key Challenges in Geography
ISBN 978-3-030-95138-2 ISBN 978-3-030-95139-9 (eBook)
https://doi.org/10.1007/978-3-030-95139-9

This Springer imprint is published by the registered company Springer Nature Switzerland AG
The registered company address is: Gewerbestrasse 11, 6330 Cham, Switzerland

Contents

Editors and Contributors

About the Editors

Theresa Bourke is an Associate Professor at the Queensland University of Technology. She has held a number of leadership positions including Academic Program Director and Course Coordinator. She teaches into a number of curriculum and discipline units, specifically in geography. Her research interests include professional standards, professionalism, accreditation processes, assessment in geographical education, and teaching about, to, and for diversity.

Reece Mills commenced his career in education as a senior secondary school geography and science teacher. He is now a Senior Lecturer at the Queensland University of Technology, Brisbane. Reece teaches into undergraduate and postgraduate initial teacher education programs in the field of primary science. His research interests include science and sustainability education as well as primary science teacher education.

Rod Lane is a former Associate Professor and Deputy Head of Department, Learning and Teaching at Macquarie University, Sydney. His research interests include educational assessment, conceptual change, and teachers' pedagogical content knowledge. Rod has published numerous peer-reviewed papers and is the author of several educational textbooks focused on secondary commerce, business studies, and geography. In 2020 Rod relocated to New Zealand to take on the position of Manager, Learning and Innovation for the New Zealand School Trustees Association. In this role he is responsible for the professional development of the country's 18,000 school board members to strengthen school governance and leadership across New Zealand.

Contributors

Theresa Bourke Queensland University of Technology, Brisbane City, QLD, Australia

Simon Catling School of Education, Harcourt Hill, Oxford Brookes University, Oxford, UK

Chew-Hung Chang Nanyang Technological University, National Institute of Education, Singapore, Singapore

Martin Hanus Faculty of Science, Charles University, Prague, Czechia

Lenka Havelková Faculty of Science, Charles University, Prague, Czechia

Deborah Heck University of the Sunshine Coast, Sunshine Coast, Australia

Injeong Jo Texas State University, San Marcos, TX, USA

Jinhee Lee Pusan National University, Busan, South Korea

Jongwon Lee Ewha Womans University, Seoul, South Korea

Sadhana Manik University of KwaZulu-Natal, Durban, South Africa

Alan Marvell University of Gloucestershire, Gloucester, UK

Rainer Mehren University of Muenster, Münster, Germany

Reece Mills Queensland University of Technology, Brisbane City, QLD, Australia

Phoebe Ow Nanyang Technological University, National Institute of Education, Singapore, Singapore

Armin Rempfler University of Teacher Education Lucerne, Lucerne, Switzerland

David Simm Bath Spa University, Bath, UK

Michael Solem Texas State University, San Marcos, TX, USA

Part I
Introduction—Surveying the Landscape of Assessment in Geography Education

Chapter 1
Binaries and Silences in Geography Education Assessment Research

Theresa Bourke and Reece Mills

Abstract This introductory chapter, written by two of the editors of this volume, surveys the landscape of assessment in geography education. After initially outlining the purposes of geography education and reviewing critical terminology with regards to the concept of assessment, the chapter that follows is divided into three sections. In Section I, a recap of the findings of a review of papers associated with assessment in geography education published between 2000 and 2016 is provided. Then, the review is updated to include publications between 2016 and the present. Analysis of this new literature in Section II reveals shifts around both assessment type and the scale of the research—what we have referred to as binaries. This section also illuminates the silences in geography education research. The chapter proceeds to argue that these binaries and silences need to be addressed if the field of assessment in geography education is to move forward in meaningful ways. Finally, Section III showcases the chapters that make up this volume, work that has already gone some way to working through the existing binaries and silences.

Keywords Formative assessment · Summative assessment · Binaries · Silences · Scale

1.1 Introduction: Geography—A Vital Subject for Citizens in All Societies

As outlined in the *International Charter on Geographical Education* (International Geography Union and Commission on Geographical Education [IGU-CGE] 2016, p. 5), 'geography is a vital subject and resource for twenty-first century citizens living in a tightly interconnected world'. According to the Charter (IGU-CGE 2016),

---

T. Bourke (✉) · R. Mills
Queensland University of Technology, Brisbane City, QLD, Australia
e-mail: theresa.bourke@qut.edu.au

R. Mills
e-mail: reece.mills@qut.edu.au

T. Bourke et al. (eds.), *Assessment in Geographical Education: An International Perspective*, Key Challenges in Geography,
https://doi.org/10.1007/978-3-030-95139-9_1

geography is the study of Earth and its natural and human environments. Often referred to as the discipline that bridges the natural and social sciences, geography is the discipline that deals with spatial variability across and between these two domains. Human activities and their interrelationships and interactions with environments from local to global scales are encouraged as essential parts of the education of all citizens in all societies. The importance of geography education lies in equipping future generations with knowledge, skills, attitudes, and practices to value, care for, and make reasoned decisions for the planet. These sentiments are equally shared by the project partners from *A Road Map for 21st Century Geography Education: Assessment* (Edelson et al. 2013, p. 7) who believed that 'geography education is essential for preparing the general population for careers, civic lives, and personal decision making in contemporary society'. These decisions should be underpinned by theoretical and applied research on geography education.

One of the research threads foregrounded in the Charter is: 'How do we understand students' learning progressions in geography? How are these best calibrated and assessed?' (IGU-CGE 2016, p. 6). It is promulgated that answering these questions allows educators to refine curriculum, pedagogic, and assessment practices used in the teaching and learning of geography. While there have been numerous books about curriculum and pedagogy in geography education, to date, assessment has not been the sole focus of a book in the discipline. The Charter equally called attention to international cooperation in the exchange of research and developments in learning and assessment. Therefore, this scholarly collection brings together research on assessment in geography education from around the globe, including Australia, Czechia, Germany, Singapore, South Korea, Switzerland, South Africa, the UK, and the USA. Before recapping and updating the geography assessment landscape from the turn of the millennium to the present to paint a picture of what is happening in this field of research, first, we overview relevant assessment terminology and position this volume in the wider global literature of assessment in education.

1.2 A Brief Outline of Assessment Literature

According to Stobart (2008), assessment can be broadly defined as 'gathering evidence in order to make a judgement' (p. 6). This aligns with Sadler's (2017) definition who posits that the primary purpose of assessment is 'to gather and distil information for reporting a conclusion or providing recommendations for decision makers' (p. 82) who can include learners, teachers, and systems. Various assessment types have been promoted in the academic literature and are now briefly elaborated.

More traditional types of assessment are usually summative tasks. Summative tasks, sometimes in the form of examinations, are used to determine the 'extent to which students have achieved intended learning outcomes' (Black and Wiliam 2018, p. 553). Sometimes referred to as 'assessment *of* learning', summative tasks occur at defined key points, for example, at the end of a unit, term, or semester; they sum up or summarize the achievement status of a student usually in the form of a rank or grade

(Sadler 1989). Their purpose is to describe learning achieved at those certain points in time as a means of reporting to parents, other teachers, or interested parties such as school boards (Harlen and James 1997). Since the nineties, summative tasks have taken the form of high stakes national or international tests that provide comparative data between states or countries, for example Trends in International Mathematics and Science Study (TIMSS) or the Programme for International Student Assessment (PISA). Results from these tests are used to inform the education community of the status of knowledge mastered by students in maths, science, or reading and are often used as a measurement of how effective teacher instruction is in these domains. Outcomes are used by policymakers as policy levers to direct, manage, and shape education systems (Riley and Torrance 2003).

This traditional view of assessment as a grade, rank, or measurement has left a legacy of assessment methods good for producing numerical values and differentiating between students' or countries' scores. However, this view has been criticized for being passive when it comes to learning and not being of much value when it comes to making judgements on understanding, application, creativity, or commitment from students (McDowell et al. 2009). This view of assessment has also been criticized in terms of the design principles that underpin assessment, that is the validity, reliability, and equity of the assessment tasks (Stobart 2008).

Validity is the alignment between the purpose of an assessment task and how well inferences are drawn from the results (Black and Wiliam 2012; Gipps and Stobart 2009; Stobart 2008). Sometimes referred to as fitness-for-purpose, close alignment between intention and outcome is required. Reliability in assessment focuses on internal consistency; that is, consistency across time, tasks, and markers (Black and Wiliam 2012; Mansell et al. 2009). It is a pre-condition for validity but is also in tension with it; attempts to increase reliability by reducing construct variance could result in under-representation of student proficiency (Black and Wiliam 2012). Equity, the last principle, is centered on fair and equitable outcomes for students. Social, economic, and political consequences of assessment, including cultural recognition and representation, and accessibility are considered (Gipps and Stobart 2009; Graham et al. 2018; Klenowski 2014). Historically, summative assessment tasks that are highly replicable advantage educational privilege (Gipps and Stobart 2009). These tasks, usually in the form of standardized external examinations, have been criticized for their limited validity and their tendency to reproduce paradigms of social inequity (Gipps and Stobart 2009; Stobart 2008). While summative assessments are still a major part of any student's education journey, there has been a big push towards using assessment as a formative endeavour.

Formative assessment has been around for over 50 years, the term often credited to Scriven (1967). A widely used definition of formative assessment accepted as a basis for good practice was developed by Sadler. According to Sadler (1989), formative assessment is focused on how teacher judgements or feedback about the quality of student responses can be used to shape and improve the student's competence. Sadler (1989, p. 121) states explicitly that 'the learner has to (a) possess a concept of the *standard* (or goal, or reference level) being aimed for, (b) compare the *actual* (or current) *level of performance* with the standard, and (c) engage in appropriate *action* which leads to some closure of the gap'. In other words, teachers must develop students' evaluative expertise so they can monitor and judge their own learning position. Formative assessment, therefore, for Sadler combines teacher feedback and student self-monitoring. Many assessment actors have defined formative assessment using various nomenclatures. These include 'assessment that supports learning' (Gibbs and Simpson 2004) and 'learning-oriented assessment' (Carless 2007). However, the phrase that has gained most traction is attributed to Caroline Gipps who was part of the Assessment Reform Group of the British Educational Research Association (BERA) in the UK in the late eighties. This group was influential in promoting the concept and practice of assessment *for* learning. A recent definition of assessment *for* learning or formative assessment which are often seen as synonymous is given below:

> An assessment functions formatively to the extent that evidence about student achievement is elicited, interpreted, and used by teachers, learners, or their peers to make decisions about the next steps in instruction that are likely to be better, or better founded, than the decisions they would have made in the absence of that evidence. (Wiliam 2018, p. 48)

Thus, assessment *for* learning usually occurs throughout the teaching and learning process to clarify student learning and understanding. In this sense, it helps students learn better, rather than just achieve a better mark. Students are provided with effective feedback that motivates them and can lead to improvement. As Willis and DeLuca (2019) put it 'collaborative power relationships are evident when the teacher physically or metaphorically sits alongside a student and through assessment processes assists a student to learn how to make value judgements and adjustments to their own learning' (p. 2). That said, teachers must ensure that Paul Black's assertion from 2006 does not ensue. Black, one of the most influential proponents of assessment *for* learning from the UK has warned that assessment *for* learning in some instances has become 'a free brand name to attach to any practice' (2006, p. 11).

Some scholars have never been truly satisfied with definitions for formative assessment and/or assessment *for* learning. This led to a third category, assessment *as* learning, entering the academic literature. Earl's (2003) interpretation of formative/assessment *for* learning, for example, saw teachers as the central characters in evaluation of student performance and providing feedback. In contrast, he believed that assessment *as* learning emphasized the students' role in self-assessment as active participants in directing their own learning. Other proponents for assessment *as* learning include, for example, Wolf (1993) and Carless (2007). However, once again this term has been criticized. For example, Sadler (2007) has argued that in assessment *as* learning 'assessment masquerades as, or substitutes for, learning itself' (p. 388) and Torrance (2007) describes it as displacing learning where achievement may be realized but a lack of understanding remains. Therefore, it appears that assessment terminology is far from straightforward and remains a site of contestation.

Regardless, as Edelson et al. (2013) argue, a focus on both formative and summative assessments 'should be integral to any efforts to improve geographic education' (p. 65). Black and Wiliam (2018) have equally argued that if classroom practice is to improve, then all aspects of assessment must be integrated in some way. Therefore, instead of seeing formative and summative assessments as binaries, they suggest that the relationship should be 'one of helpful overlap—and perhaps mutual support' (p. 554).

This chapter that we have aptly titled *Binaries and Silences in Geography Education Assessment Research*, is divided into three sections. In Section I, we overview the literature in geography education assessment from 2000 to present. This is achieved by, first, recapping the findings from Lane and Bourke's (2019) systematic review published in *International Research in Geographical and Environmental Education* (IRGEE) and, second, by updating the geography assessment literature from the last six years, replicating the search from the 2000–2016 systematic review. The analysis that proceeds from this in Section II shows shifts in the field related to binaries in scale (local/global) and assessment types (formative/summative) and silences in the research literature that we argue must be addressed to move the field forward in meaningful ways. We use these binaries and silences in Section III to discuss chapters from the present volume to identify examples where we see authors realizing these aims. In doing so, we outline the structure of this volume wherein the following 10 chapters are divided into four parts based on themes we generated through a careful reading and curating of the contributions.

1.3 Section I: Recapping and Updating the Research Literature

This part of the chapter first recaps the findings of a review on geography assessment written by two of the book co-editors, Rod Lane and Terri Bourke, called *Assessment in Geography Education: A Systematic Review* published online in 2017 and then in print in 2019. In this review, the authors examined the international academic literature on geography assessment practices between 2000 and 2016. Thirty studies were shown to meet the selection criteria and were categorized into the following themes: (1) formative assessment; (2) spatial reasoning; (3) performance/achievement standards; (4) marking; (5) content analysis of assessment items (alignment with national policy requirements); (6) international assessment; (7) other assessment strategies (free word association, sketch mapping, self-evaluation); and (8) case studies and evaluations of national practice.

In this systematic review, 12 countries were represented, including the US, the UK, Australia, Bulgaria, Turkey, Germany, Spain, the Netherlands, Zimbabwe, Rwanda, New Zealand, and India. Most of the studies were either national or local in scale, with only four crossing international boundaries. Studies were conducted with secondary school students; only two were undertaken in primary school settings. Two studies looked across primary and secondary contexts and three across secondary and tertiary. A range of designs were present including document analysis, experimental design, interviews and observations, policy analysis, think aloud protocol, case study, and questionnaires. Broadly, the studies were classified as qualitative (53%), quantitative

(18%), and mixed methods (29%) with participant numbers ranging from ten to 2,879 students/teachers.

While the authors acknowledged that numerous discussion papers and reports in geography education, for example *A Road Map for 21st Century Geography Education: Assessment* (Edelson et al. 2013), have contributed to our understanding of the general principles of assessment design, the review found that there remains a lack of evidence-based practice and empirical research on the topic. The authors concluded that there was an absence of longitudinal research providing ongoing evidence of the impact of assessment on learning outcomes. Most studies were small-scale, either local or national, mainly in the UK or US. There was also a lack of connection between researchers examining the same problem or questions in different locations or contexts, that is, they did not cite each other. In sum, there was no coherent direction or collaboration on a global scale. There was also a distinct lack of research in assessment located in the primary sector as well as valid and reliable assessment instruments for measuring the attainment of key geographical concepts and skills. Issues of alignment were also highlighted, where assessment practices had not always adjusted to changes in education standards and curriculum documents. Furthermore, there was a paucity of studies examining progression in geographical understanding. The authors called for studies located in primary education, mapping of learning progressions—the essential knowledge and skills students in geography should develop over time—as well as instruments to provide information on this progress. The chapters in this volume go some way to addressing these gaps and will be introduced in Section III.

The purpose of the next section of this chapter is to update the original systematic review reported on above to take into consideration studies about assessment in geography education from 2016 to 2021. In replicating the original search parameters and after abstract and full text screening, only eight studies were found (Table 1.1). The findings from the updated literature search revealed two distinct themes: (1) standardized testing ($n = 6$); and (2) instrument development ($n = 2$). The studies are first described in terms of their general characteristics (place of research, participants, and methodology).

Table 1.1 Studies in the updated review 2016–2021

Standardized testing	Instrument development
Baird et al. (2016)	Cox et al. (2018)
Bijsterbosch (2018)	Nursa'ban et al. (2020)
Bijsterbosch et al. (2016)	
Eze (2021)	
Fjellborg and Kramming (2021)	
Solem et al. (2018)	

1.3.1 *Findings—2016–2021*

The following countries were represented in the new set of included studies—Holland ($n = 2$) and one each from Ireland, Sweden, Nigeria, Indonesia, and Belgium. Also included was one international study, spanning across several countries. The studies were classified as qualitative (25%), quantitative (25%), and mixed methods or multi-methods (50%), with participant numbers in the quantitative and mixed/multi-methods studies ranging from 595 to 1,002 students/teachers. The qualitative studies had no participants as they comprised content analysis of national geography examinations/chief examiners' reports. The two themes are now elaborated.

1.3.1.1 Standardized Testing

There was one study by Solem and colleagues (2018) that focused on standardized assessment at a global level. This study reported on the schedule of a group of international academics working towards framing an international assessment in geography. An evidence-centered design (ECD) was outlined consisting of five components: domain analysis, domain modelling, assessment framework, assessment implementation, and assessment delivery. This paper specifically reported on the first two where a survey, document and instrument analysis, and a social lab were the chosen data collection and analysis methods. The survey was conducted with the seven member countries of Trends in Geography Assessment Study (TIGAS) which include Australia, USA, The Netherlands, South Korea, Switzerland, Singapore, and Czechia. Participants were asked questions about the geography curriculum in their country for 13–14 year olds. Findings from the survey and document analysis revealed that 75% of the countries surveyed offered geography as a stand-alone subject with a high percentage of common geographical issues, domains, skills, and concepts present. The group also analyzed geography assessments from Singapore, the USA, and Australia to categorize them according to targeted ability, item characteristics, and confounding factors. For this part of the study, a comparative matrix was developed, outlining for example the item type, stimuli, topic areas, setting, instructions, and structure. This process enabled the team to realize how this could happen at a much larger scale. From these findings the provisional framework was developed, and feedback was sought through a social lab. These findings will inform the next iteration of the framework going forward.

Five studies focused on standardized assessment at the national level (Baird et al. 2016; Bijsterbosch 2018; Bijsterbosch et al. 2016; Eze 2021; Fjellborg and Kramming 2021). In the study by Baird et al. (2016), students' views on the predictability of the Higher-Level Irish Leaving Certificate (LC) were investigated to determine if there was any relationship between predictability and examination outcomes. Data were collected from surveys with 1,002 students taking examinations in biology ($n = 536$), English ($n = 749$), and geography ($n = 387$) either online or by pencil and paper. This quantitative study used regression analysis to determine correlations between perceptions of examination predictability (measured on Likert-scales) and students' scores on the examination. Findings for geography showed that the scoring system rewarded students who felt that they should study their discipline broadly and have a certain degree of test-wiseness. Narrowing the curriculum in preparation for the test in general tended to lower outcomes. The authors pointed out that these findings could inform teacher practice but warned about generalizability.

In the next study, Bijsterbosch (2018) conducted an analysis of national geography examinations in the Netherlands (2015–2017) to determine their content and alignment to the local geography curriculum. In doing so, examinations were analyzed using Bloom's revised taxonomy (Anderson and Krathwohl 2001) for geographical knowledge and cognitive dimensions. Bijsterbosch (2018) found that most test items focused on recall of knowledge, with higher-order cognitive processes such as evaluating and creating completely absent. He concluded that the national examinations do not assess higher-order thinking skills possibly due to the dominance of multiple-choice and short answer questions. A greater emphasis on constructive alignment was called for, with the author recommending assessment practices more aligned with the broad purposes of geography education such as students' knowledge and understanding of the world and their relationship with it.

A further study, also located in the Netherlands by Bijsterbosch et al. (2016), examined senior secondary geography teachers' assessment practices. Using a survey of multiple choice and open-ended questions, the participants ($n = 74$) were asked about their practices related to school-based summative examinations. The main finding was that geography teachers in this study primarily sourced test items from textbooks or previous examination papers, with a limited proportion constructing test questions themselves. Bijsterbosch et al. (2016) proposed that standardized national external examinations were influencing teachers' assessment practices with respect to context and format of test items. Further research and professional learning for teachers in this area were suggested by the authors as ways forward.

Eze (2021) used a qualitative grounded theory approach to analyze ten years (2008–2018) of chief examiners' reports from the West African Examination Council (WAEC) to determine why Nigerian students perform poorly in geography examinations. There were three dominant reasons for this including inadequate preparation, inadequate explanations, and poor mapwork. Additionally, poor comprehension of questions was seen as a factor as well as poor expression in the presentation of answers. Furthermore, students often did not adhere to the rubrics provided.

The final study in this theme examined Swedish national geography assessment for gendered response differences between boys and girls (Fjellborg and Kramming 2021). The authors' study focused on questions related to sustainable development as a focal point of geography education and the Swedish national curriculum. Ninth grade students' test responses from years 2014 to 2019 ($n = 1{,}500$) were analyzed using several quantitative techniques. Overall, it was found that girls performed better than boys on tests. Of significance were gendered trends related to the response format of test items. Girls performed better on short/long- response items (constructed response items), whereas boys performed better on multiple-choice questions. This demonstrated that girls performed better on items requiring higher cognitive work such as synthesis, analysis, and evaluation. The authors suggested attention to inclusive assessment designs and diverse conceptualizations of sustainable development to engage both girls and boys.

1.3.1.2 Instrument Development

The theme of Instrument Development was focused on the testing and measurement of systems thinking in geography education at regional ($n = 1$) and national ($n = 1$) scales. The first paper by Cox et al. (2018) reported on the development of a test instrument to measure Flemish secondary school students' (16 to 18 years old; $n = 617$) abilities to understand interconnections between geographical ideas. The instrument comprised six short-response items based around cause and effect scenarios. While the authors argued the test as a valid and reliable way to measure systems thinking, they called attention to the challenge of operationalizing systems thinking as a construct to be measured. They noted the importance of students being given opportunities to think about cause and effect across local, regional, and global scales.

The final study, by Nursa'ban et al. (2020), outlined the development of a measurement instrument for spatial thinking in geography. Spatial thinking was argued to be an important part of geography education, connecting social and physical dimensions. Participants in this study were a mix of academics, school leadership (e.g., principals), teachers, and students ($n = 595$) located in Indonesia, who completed interviews and a questionnaire. The authors conducted a Confirmatory Factor Analysis, among other statistical procedures, to determine a 15-factor model of spatial thinking. This foundational work appears to be informing geography spatial thinking in Indonesia.

From these findings, it can be concluded that there remains a paucity of research into assessment in geography education in general. Only eight studies across the timeframe met the inclusion criteria with all studies either focused on standardized testing or instrument development. With most studies focused at the national scale and geared towards assessment *of* learning, the binaries pointed out at the beginning of this chapter are reflected in this updated academic literature. The next section, *Cross Analysis to Provide New Insights*, further elucidates the binaries and silences.

1.4 Section II: Cross Analysis to Provide New Insights: Binaries and Silences in Geographical Assessment Research

In this section of the chapter, we analyze the shifts in research about assessment in geography education. To achieve this aim, we located the studies from Section I of this chapter in a matrix according to scale (local/regional, national, international) and assessment type (assessment *for*/*of* learning). This allowed us to track how research trends have shifted from 2000 to present. We made decisions about where to locate research studies through a reflexive and collaborative process that primarily relied upon our professional judgement as geography educators and teacher education researchers. We also used definitions related to scale and assessment type. For definitions of scale, we used the Oxford English Dictionary. These definitions are:

- Local/regional = belonging to or connected with a particular place or area, for example, one school or classroom in a neighborhood or region;
- National = connected with a particular nation; shared by a whole nation, for example 'representative' sample or from multiple sites across national borders (different states/counties);
- Global = covering or affecting the whole world, for example across international borders.

Definitions for assessment types were outlined at the start of the chapter. If studies could be categorized as both 'assessment *for* learning' and 'assessment *of* learning', the authors names spanned both the columns of the table to show this (see Table 1.2). Some studies could not be categorized according to assessment type for various reasons (referred to in Table 1.2 as non-applicable), for example three studies analyzed questions from geography textbooks that could have formative or summative purposes (Mishra 2015; Yang 2013; Yasar 2009) and one study analyzed curriculum and assessment guidelines broadly (Anthamatten 2010).

The first trend observed in the data showed studies shifting from local/regional to national in scale. For studies between 2000 and 2016, four were located globally (13%), nine studies were located nationally (30%), and 17 were located at the local/regional scale (57%). For studies between 2016 and 2021, one was located globally (13%) and the remaining seven were national (87%), with no studies at the local/regional scale.

The second trend observed in the data reveals that studies have shifted from assessment *for* learning to assessment *of* learning. Between 2000 and 2016, seven studies focused on assessment *for* learning (23%) and 15 focused on assessment *of* learning (50%). Two studies spanned the boundary of assessment *for* learning and assessment *of* learning (7%). The other six studies between 2000 and 2016 that we have termed non-applicable comprise ones that could neither be categorized as assessment *for* learning nor assessment *of* learning (20%). For studies 2016–2021, there are no studies about assessment *for* learning. All studies (100%) are related to assessment *of* learning.

Table 1.2 Matrix showing shifts in research about assessment in geography education (2000–2021)

	Assessment *for* learning	Assessment *of* learning	Non-applicable
Global		**Solem et al.** (2018) Bourke and Lane (2017)* Lane and Bourke (2017)* van der Schee and Kolkman (2010)	Wertheim et al. (2013)
National		**Baird et al.** (2016) **Bijsterbosch** (2018) **Bijsterbosch et al.** (2016) **Cox et al.** (2018) **Eze** (2021) **Fjellborg and Kramming** (2021) **Nursa'ban et al.** (2020) Bijsterbosch et al. (2017)* Crisp (2008) Davies (2002) Pinar (2011)	Anthamatten (2010) Mishra (2015) Yang (2013) Yasar (2009)
	Butt et al. (2006)		
Local/ regional	Baldwin (2003) Budke et al. (2010) Cooper and Cowie (2010) Davies et al. (2004) Leat and Nichols (2000) Panadero et al. (2012) Tiknaz and Sutton (2006)	Andrews et al. (2008) Atasoy (2009) Battersby et al. (2006) Harwood and Rawlings (2001) Huynh and Sharpe (2013) Lee and Bednarz (2012) Munowenyu (2007) Tomaszewski et al. (2015)	Lock and Barrett (2002)
	Wehry et al. (2012)		

Not Bold: 2000–2016 **Bold:** 2016–present.
*Published online in the timeline for the original 2000–2016 review

The third trend relates to the numbers of studies throughout the last 21 years that have crossed boundaries related to assessment type, that is, connecting assessment *for* and *of* learning. Between 2000 and 2016, only two of 30 studies (7%) spanned assessment type boundaries and from 2016 to 2021, no studies fell into this category.

These trends reveal that there remains binaries in scale of study, that is, local versus global and assessment type, that is, assessment *for* versus assessment *of* learning. Before trying to explain these binaries and provide recommendations for future work, our survey of the landscape of assessment in geography has also exposed numerous silences across the corpus of research that we think are noteworthy for future exploration:

- There remains a lack of research in assessment related to cross-cutting concepts and skills in geography education, for example, learning progressions.

- While spatial thinking was a dominant theme in the original review (2000–2016), only one study progressed this research in the updated review (2016–2021).
- In the updated systematic review, Cox et al. (2018) put forward a valid and reliable instrument for systems thinking; however, they outlined many challenges in operationalizing such a measure. They called for more research into cause and effect in systems thinking across scales.
- There was a complete lack of research related to assessment of two of the signature pedagogies of geography, inquiry, and fieldwork.
- Despite some attention to research related to validity and reliability in the last two decades, there is a notable absence of research attending to the fundamental principle of equity, that is, inclusive assessment design and practices for all learners.
- There is a noteworthy absence of assessment research as a social justice endeavour, that is, how assessment data can be used to address disadvantaged learners.
- The research corpus is responding to curriculum reform and global education discourses with no creative, innovative, or technological-driven approaches to assessment in geography.
- There is a dearth of research related to classroom/school-based assessment practices, for example, design of task sheets and marking rubrics as well as moderation activities.
- In the 2000–2016 systematic review, assessment in primary school geography was earmarked as an under-researched area. This remains the case for literature 2016–2021, with no studies located in this school context.

1.4.1 Discussion and Recommendations for Future Research

As mentioned previously, in the late eighties, the Assessment Reform Group from the UK with assessment actors such as Caroline Gipps, Paul Black, and Dylan Wiliam led a global charge on assessment *for* learning practices. From the analysis of papers in the geography assessment academic literature between 2000 and 2016, assessment *for* learning was found to be a dominant theme. This could infer that geography education researchers were trialing and testing assessment *for* learning practices in line with the assertions from the Assessment Reform Group. Examples included Baldwin's study (2003) about the role of feedback, or self-assessment practices in Davies et al. (2004). Some of the reported benefits were enhanced student motivation and learner autonomy (Cooper and Cowie 2010). Additionally, if students explicitly understood assessment criteria, they could make more sense of teacher feedback to improve their work (Davies et al. 2004; Leat and Nichols 2000).

However, at the turn of the century and originating in science and maths, there was an onset of global standardized assessment *of* learning. This shift was in response to the quality agenda in education where teacher and teaching quality, professional standards, accountability, and performativity became the dominant prevailing discourses (Bourke et al. 2016). Global tests, such as Trends in International Mathematics and Science Study (TIMSS), first implemented in 1995 for maths and science, and Programme for International Student Assessment (PISA), first administered in 2000 for maths, science, and reading, emerged. These tests provide comparative data, in the case of PISA for over 80 countries whilst the TIMSS assessment performs a unique role by providing a snapshot of the performance of cohorts across more than 60 countries. In this era of accountability, international surveys of educational achievement have become commonplace despite being contested in relation to issues of validity, reliability, and equity.

The global standardized assessment *of* learning sentiment entered geography several years later. In 2013, as part of the Road Map project, assessment was pointed out as an area that needed attention. The partners in the Road Map, the American Geographical Society (AGS), the Association of American Geographers (AAG), the National Council for Geographic Education (NCGE), and the National Geographic Society (NGS), not only recognized assessment results as indicators of educational progress, but also believed in their utility for 'decision making in education policy and practice' (Edelson et al. 2013, p. 7). The Road Map authors argued that assessments could measure students' proficiency against benchmarks, inform decisions about the function and effectiveness of programs, evaluate class/school performances, and be used to focus resources and set research directions. Recommendations from this report included the continuation of the National Assessment of Educational Progress (NAEP), but a revamped version, so that items went beyond testing declarative knowledge to other geographic practices.

In 2014, some of these initial murmurings about larger scale summative assessments sparked an editorial in *International Research in Geographical and Environmental Education* by the editors at the time, Professors Joe Stoltman, John Lidstone, and Associate Professor Gillian Kidman. The editors called attention to the fact that geography was not on the international assessment map and they suggested that a coordinated effort should be made by academic and professional associations to advocate for the inclusion of geography in TIMMS (Stoltman et al. 2014). In the following year at a meeting at the annual conference of the IGU-CGE in Moscow, the conversation around an international assessment in geography continued. It is perhaps in response to this momentum that the studies from 2016 to 2021 solely focused on assessment *of* learning, albeit mostly at a national scale. For example, Bijsterbosch's work (2018) examines national geography assessment in the Netherlands and how national external examinations influence teachers' assessment practices with respect to test format and the absence of cognitive levels beyond knowing and recall. Other authors have investigated how test question constructs can favor particular genders. For example, Fjellborg and Kramming (2021) found that girls performed better than boys on short and extended response items. Additionally, Eze (2021), working in a West African context, illuminated how students' geography results on summative tests were hindered by low levels of comprehension. While Bijsterbosch's study brings into question issues of validity, the latter two studies (Fjellborg and Kramming 2021; Eze 2021) point to equity issues, two principles underpinning effective assessment design.

Another possible catalyst explaining the shifts in the findings from the literature related to scale and assessment type was Lane and Bourke's (2019) systematic literature review on assessment in geography published in IRGEE. As mentioned previously, the review called for more global efforts in understanding geography assessment, in particular connecting similar programs of research conducted in different locations. One of the recommendations from the literature review was for 'large-scale studies … with international researchers working collaboratively on an agreed set of research questions' (p. 33) such as 'an international assessment of geographical literacy' (p. 34). The work of one group of researchers—Trends in International Geography Assessment Study (TIGAS)—responded to this call by championing the possible inclusion of geography in TIMSS (e.g., Bourke and Lane 2017; Lane and Bourke 2017; Solem et al. 2018). This research group, whose studies are represented in the updated literature review (Section II), examined stakeholders' perceptions about an international geography assessment (Bourke and Lane 2017; Lane and Bourke 2017) as well as proposing a framework and methodology to guide the specifications for test items, evaluation procedures, and measurement models (Solem et al. 2018).

Proponents for large-scale standardized assessment maintain that these data are valuable for the decision making of multiple stakeholders in education in terms of policy and practice (Baker and Wiseman 2009; Brown et al. 2005; Koretz 2009). Cai and Lester (2007) additionally highlight how these data can be used to inform policy and practices in response to the needs of specific groups of children such as those from diverse ethnic backgrounds. However, other authors such as Wrigley (2004) point out that international testing is reductionist, often masking individual or regional-level variations. Prais (2007) concurs with Wrigley's criticisms; skeptical about coverage, sampling, administration, and interpretation of such large-scale data. Furthermore, often countries are ranked against each other without considering the differences between the compared countries.

While the focused effort in understanding standardized national and international testing certainly responds to the calls for such research, what our analysis has revealed is that this push for assessment *of* learning has been at the expense of studies in assessment *for* learning, which has received little to no attention in the literature in the past six years (2016–2021). Proponents of assessment *for* learning argue that this type of assessment is non-evaluative and supportive of the learner. Timely feedback, a central tenet of assessment *for* learning, encourages students to be engaged, motivated, and to self-monitor to improve their outcomes (Cooper and Cowie 2010). Other authors, however, have indicated this type of assessment is poorly understood by teachers, time consuming, and as mentioned earlier has become 'a free brand name to attach to any practice' (Black 2006, p. 11).

Whilst acknowledging strengths and limitations of different types of assessment and the scale at which they are administered, we argue there appears to be an artificial dichotomy related to notions of (1) scale and (2) assessment type, wherein local/global or assessment *for* learning/assessment *of* learning are often represented as opposed or separate from each other. We suggest that these are two binaries (ends of a continuum) that must be worked *through* because efforts at understanding all aspects of assessment are more valuable than understanding each individual type alone (e.g., Black and Wiliam 2018). In regard to working through these binaries, questions that we see in need of further research in assessment in geography education are:

- What synergies exist between formative and summative assessment types?
- How can assessment design integrate formative and summative assessment types?
- How can formative assessment contribute to the purposes of summative assessment?
- How can data from large-scale standardized assessment be used by students and teachers in learning and instruction?
- How can formative and summative assessment types be located in signature pedagogies such as inquiry-based learning?

We encourage these questions to be investigated at local/regional, national, and global scales or span different or multiple scales.

In regard to working through the silences in the literature, programs of research that we see in need of further attention in assessment in geography education are:

1. Assessment related to cross-cutting concepts and skills in geography
2. Assessment related to the signature pedagogies of inquiry and fieldwork
3. Assessment as a social justice endeavour in geography education
4. Future directions for assessment in geography education.

We encourage researchers to adopt an agentic stance, being proactive in setting future directions in research wherein geography educators decide what to research instead of being steered by curriculum and assessment reforms as part of broader quality, accountability, and performativity regimes in education. The chapters included in this volume take up these programs of research, with 10 chapters structured into the four sections listed above, each making a substantive contribution to progressing research in geography education assessment.

1.5 Section III: Structure of this Volume

This section of the chapter describes how the proceeding volume is structured and outlines each contribution.

1.5.1 Part I: Assessment Related to Cross-Cutting Concepts and Skills in Geography

The three chapters in Part I collectively explore assessing fundamental building blocks or cross-cutting concepts/skills in geography education, namely systems thinking, mapping, and spatial thinking. These chapters with writers from Germany and Switzerland (Rainer Mehren and Armin Rempfler) and Korea and the USA (Jinhee Lee and Injeong Jo, and Jongwon Lee and Injeong Jo) based in German and Korean contexts are all empirical studies. The first two specifically focus on the validation of assessment tools in geography education and the final study is a systematic review of assessment tools used in geography for spatial thinking. The common thread across these chapters indicates the need for theoretically and empirically derived assessment tools that can be used both formatively and summatively to measure cross-cutting concepts and skills in geography that are largely silenced in research. It is only by having such tools that learning, teaching, and assessing in the subject will improve.

The first chapter, *Assessing Systems Thinking in Geography* by Rainer Mehren and Armin Rempfler, presents an in-depth quantitative study where an empirically validated tool for assessing systems thinking in geography is tested. The purpose of this tool is to define what competency and proficiency in systems thinking looks like and how it could be promoted in practice. The authors present varied examples of tested items based on the topic of atmospheric systems: multiple choice, short answer, and extended responses including concept mapping questions. As the authors state, 'systems thinking or system competence is considered a fundamental concept' of the school subject of geography in Germany. Systems thinkers take into account 'cognitive analysis and mental representations of geographical phenomena', providing deeper understandings of the great challenges of the twenty-first century such as climate change and socio-economic disparities between North and South. The authors make several useful recommendations for the effective teaching of systems thinking such as starting in the early years, integrating it in a spiral-curricular design where content across lessons and topics are connected, using graphical representations including concept mapping and computer simulations, and allowing students to think about problems from multiple perspectives. The authors' contribution is important as having a theoretically derived and nationally validated measuring instrument for classroom teaching interventions to promote systems thinking is vital to geography education.

The aim of the second chapter in Part I, *Assessment of Mapping Learning Progressions in Geography* by Jinhee Lee and Injeong Jo, is twofold. First, the authors provide methods for examining learning progressions (LP) in geography, outlining in detail three approaches: data grounded, in-situ, and instructional interventions. And, second, they outline a process for measuring the validity of learning progressions in mapping, an emerging research field in geography education. Responding to the call by the influential *Road Map for 21st Century Geography Education* Project, to engage in systematic efforts to identify learning progressions in geography, the authors set out to determine appropriate curriculum scope and sequences aligned with students' developmental levels in specific mapping concepts including contour lines, distance, location, map projection, orientation, and scale. Using Corcoran et al.'s (2009, p. 15) definition, learning progressions are 'empirically-grounded and testable hypotheses about how students' understanding of, and ability to use, core scientific concepts and explanations and related scientific practices grow and become more sophisticated over time, with appropriate instruction'. The authors meticulously outline the data collection and analysis steps they took to develop a hypothesized LP and then pilot and test an assessment to obtain information about what students know and are able to do at different levels. In this Korean study, the authors demonstrate the powerful role assessment plays in informing geography teaching and, in their words, 'identifying whether a curriculum is aligned with students' developmental stage (LP)' and 'whether the learning objectives are appropriate and whether the content and processes proposed to achieve the learning objectives are suitable'.

The final chapter, *Assessing Spatial Skills/Thinking in Geography* by Jongwon Lee and Injeong Jo presents a synthesis of peer-reviewed studies on the measurement/assessment of spatial skills/thinking. Informed by systematic review protocols, 19 studies made up the final corpus of papers for full analysis. The impetus for the chapter was in response to the assertion from the National Research Council's (2006, p. 232) report *Learning to Think Spatially* that claimed that there were not 'valid and reliable assessments for spatial thinking'. After outlining the conceptualizations and various emphases across disciplines for spatial thinking, the methods for the review including search strategy and data extraction and analysis are outlined. The authors discovered four broad trends—measuring spatial skills/thinking on a small scale and without a specific context; measuring spatial skills/thinking on an environmental scale and in everyday contexts; measuring spatial skills/thinking on a geographic scale and in geoscience contexts; and measuring spatial skills/thinking self-assessments in everyday life contexts. These broad trends correspond with the authors' sentiment in the opening line of the abstract where they say that 'The acquisition of spatial skills/thinking is essential for both everyday functioning and academic purposes'. Despite the detailed analysis from the authors, they concluded that research on tools for measuring/assessing spatial thinking is still limited with inherent problems of contested meanings across disciplines. The authors called for interdisciplinary understandings to further develop spatial thinking measurement research.

1.5.2 Part II: Assessment Related to the Signature Pedagogies of Inquiry and Fieldwork

Part II of this volume concentrates on the signature pedagogies of inquiry and fieldwork as it relates to assessment, with contributions from the UK (Alan Marvell and David Simm) and Singapore (Chew-Hung Chang and Phoebe Ow). Although variously described by the authors as 'the heart of geography' (Marvell and Simm) and a 'hallmark of geographical education' (Chang and Ow), the common message about inquiry and fieldwork in these chapters is the importance of including formative assessment (assessment *for* and *as* learning) as well as the more traditional summative task (assessment *of* learning). Both these chapters show as Black and Wiliam (2018) put it 'helpful overlap' or 'mutual support' (p. 554) between formative and summative approaches.

The first chapter, titled *Integrating Assessment Effectively into International Fieldwork: A Case Study Using Student-Led Teaching and Learning* written by Alan Marvell and David Simm, showcases an international fieldwork experience for geography and tourism management students at Bath Spa University, UK. In this chapter, Marvell and Simm depict a 'transformative learning experience' where students are immersed in an unfamiliar, international environment (Barcelona) which 'challenges their preconceptions' and invites affective emotional responses which in turn lead to a 'deeper learning experience'. The main argument is that students should be formatively and summatively assessed through multiple stages of a field trip. The summative assessments in this Advanced Geographical Fieldtrip module include a group project proposal, a group presentation, an individual fieldwork diary, and two essays, one reflective in genre. However, these summative tasks also encompass formative feedback channels incorporating peer/tutor–group discussions and debate, reflective fieldwork diaries, peer–peer learning, and co-production of knowledge and understanding as ideas are shared in a community of learners. The authors highlight fieldwork as an effective pedagogy for 'critical thinking, problem solving, communication, and leadership as well as collaboration, adaptability and curiosity … and interviewing and negotiation skills', all useful for future studies, life, and employability. This chapter particularly illuminates integration of assessment types and working through the binary of formative and summative assessments at a global scale.

In the second chapter, *Inquiry-Based Fieldwork Assessment* for *and* as *Learning in Geography,* Chew-Hung Chang and Phoebe Ow posit that 'adopting an inquiry-based approach to assessment in an authentic setting not only enables students to develop deeper geographical understanding, but offers a modality of assessment for us to know if students have learnt what was intended along the way'. In other words, these authors highlight that inquiry based pedagogies and formative assessment in fieldwork are inextricably linked. In their opinion, assessment in fieldwork should not be limited to summative tasks such as a final report or presentation. The authors bring their arguments to life by applying their thinking to a Sustainability Learning Lab (SLL) in Singapore. Reflexivity and feedback are emphasized as important aspects of inquiry and assessment in this virtual laboratory. Pre-service teachers are prepared to teach fieldwork components outlined in the local official curriculum. They learn how to 'use fieldwork equipment and what data are best to collect, interpret, and analyze to answer the inquiry question'. A double loop of reflexivity for active learning is enabled by self and peer assessments including, for example, annotating photographs with reflections of personal experiences. All of these formative strategies help with the final product—the summative assessment task—and allows pre-service teachers to gain confidence for their future teaching. Once again, this chapter shows integration and working through the binary of formative and summative assessments.

1.5.3 Part III: Assessment as a Social Justice Endeavour in Geography Education

The chapters in Part III of the book view assessment in geography education from a social justice perspective. This part comprises two chapters. These chapters are from the US (Michael Solem) and South Africa (Sadhana Manik). These chapters present contrasting implications related to *assessment of learning* in terms of equity. While Solem's work shows the power of large-scale standardized assessment to illuminate disadvantages for specific groups of learners, Manik's work shows how, despite the best of intentions, inequalities can be exacerbated where assessments require funding, trained teachers in GIS, and technological resources that the poorest schools cannot access.

The first chapter, *The National Assessment of Educational Progress (NAEP): A Closer Look at US Gaps and Trends in Geography Achievement* written by Michael Solem, overviews the NAEP assessment programme conducted in the US before examining geography data related to race and ethnicity, gender, and socio-economic status. In this chapter, Solem paints an alarming picture of 'disparities' and 'deprivation', wherein certain groups—Black, Hispanic, and multiracial students, female students, and lower socio-economic status students—experience inequitable educational disadvantage. He emphasizes the importance of geographical knowledge for active participation in society and challenges us to respond to the 'systemic denial of the "pedagogic rights" of all young people to access powerful [geographical] knowledge'. Although sobering, Solem's work highlights the value of large-scale standardized assessment for understanding learning trends in geography education from a social justice perspective. He concludes the chapter by acknowledging the role of large-scale data sets such as those provided by the NAEP assessment programme in answering research questions that can meaningfully impact policy and practice and ultimately '[offer] a path toward a better future'.

The second chapter, *Focusing on Quality, Forgetting Inequalities: Assessment Within GIS in the Geography Curriculum and Assessment Policy Statement (CAPS) in South Africa* written by Sadhana Manik, begins with a mapping of national curriculum and assessment reform in South Africa since apartheid. In doing this work, Manik points out how several iterations of policy have attempted to remedy the racism explicit in the national education system with varying success. She then focuses her analysis on the way that Geographic Information Systems (GIS) are incorporated in the latest iteration of the Grade 10–12 geography curriculum, noting, for example, its inclusion in high-stakes summative assessment tasks that have substantive implications for students' further education, training, and/or work. Manik problematizes this finding, saying 'disadvantaged schools often select the GIS essay option ("learning about GIS") whilst advantaged schools are propelled towards the GIS research option ("learning through GIS"). Thus, maintaining the same level of quality for all learners in GIS teaching and assessment is clearly lacking'. She further demonstrates how this racial inequality, enabled through GIS assessments, is exacerbated by a lack of teacher preparedness to teach GIS in predominantly Black public schools. Manik's critique illuminates how assessment can be used as a socio-political tool, which in the South African context is working against the realization of social justice. She leaves us with the image of the 'inequalities prevalent in disadvantaged public schools and the disparity in resources needed (tools and teacher training) to ensure quality education for all learners including learners who are poor, Black, and bounded by the vestiges of an apartheid history'. It is through parity in resources, professional learning, and integration of assessment *for* learning and *of* learning, Manik argues, that this curricula space will be transformed to more fertile terrain. This is yet another example of integrating different types of assessments, in this case for issues of equity.

1.5.4 Part IV: Future Directions for Assessment in Geography Education

The three chapters in Part IV conclude the volume by considering future directions for geography education assessment. These chapters are from Czechia (Martin Hanus and Lenka Havelková), Australia (Deborah Heck), and the UK (Simon Catling) and respond to a number of the silences in research in geography education alluded to earlier.

The first chapter, *Online Adaptive Assessment of Map Skills* written by Martin Hanus and Lenka Havelková, provides insight into how digital technology in geography education has allowed educators to think differently about assessment. Hanus and Havelková inform us how assessment has pluralistic purposes in contemporary education and present a compelling argument that computer-based online adaptive assessment is a means to achieve feedback and self-regulation in learning. They define adaptive assessment as 'personalized … assessment according to the actual [learning] position of each student', wherein each student answers adaptive questions 'according to his/her previous success'. In this chapter, Hanus and Havelková illustrate this definition through showcasing an application called Mapwork.education. These authors alert us to future possibilities where online adaptive assessment may be used in an integrated way as both assessment *of* learning and assessment *for* learning with a choice in format between right/wrong answer or students receiving instruction in the form of 'hints' that scaffold them to the correct answer. Possibilities also exist to 'make data-driven decisions in the classroom' around students' pre-instruction conceptions, learning progression, and instructional needs. Hanus and Havelková quite rightly conclude the chapter by asserting the need for investment in this future direction for assessment in geography education in the form of technical programming knowledge, digital systems, and professional learning for teachers.

The second chapter, *Engaging with Conceptual Acrobatics: Geography Assessment Standards and Rubrics* written by Deborah Heck, envisions a future wherein assessment tasks and marking rubrics are co-constructed with learners. Set against the backdrop of accountability and audit discourses in education, Heck positions rubrics as tools that can work to reinforce or challenge 'systematic approaches to measure performance and outcomes' in education. Her analysis highlights how rubrics may employ representational, sociocultural, or sociomaterial discourses related to standards, and she illustrates her point with examples from the geography components of the Australian curriculum. Heck illustrates how rubrics can be more than a static 'checklist' by arguing for reflexivity in the design of dynamic and adaptable criteria. This allows for divergent student responses in assessment. To achieve this aim, she calls for geography educators to 'consider ... their own conceptions of assessment' and 'reconstruct their own identity as an assessor in the geography classroom'.

The final chapter in this much-needed visionary section, *Reflecting on the Assessment Landscape of Primary Geography* written by Simon Catling, takes a deep dive into primary geography education, where there is a monumental dearth of assessment research. Catling asserts that there has been thoughtful and useful guidance on the topic; however, he repeats the point quite a few times that assessment in primary geography is an area ripe for research. He provides various avenues that could be pursued for research, outlining many approaches used in geography classrooms which are yet to be tested to see their worthwhileness or impact. Some examples include: *How do primary teachers approach assessment during their geography lessons and topics?* and *In what ways does geography assessment contribute to the evaluation and improvement of geographical provision, teaching, and learning and their quality across a school?* Catling sees assessment as tied explicitly to learning in geography and summarizes his chapter by claiming 'the value of assessment research must be to develop ever more effectively children's learning in and of geography and teachers' capability to provide for this'.

1.6 Concluding Remarks

This chapter has surveyed the landscape of assessment in geography education from the turn of the millennium to the present moment. *Binaries* in both scale and assessment types were foregrounded as well as the *silences* in the academic research literature on geography education assessment. We call for geography educators to work through these *binaries* including research around both international/national high-stakes assessment and localized classroom-based assessment practices and how assessment *for* and *of* learning could speak to each other across scales. We further call for proactive research that addresses the *silences* in scholarship around cross-cutting concepts and skills, signature pedagogies including learners' emotional experiences of assessment, inclusive assessment practices for diverse learners, innovative formats/genres as assessment tools such as adaptive online assessment, and geography assessment in the primary school. As articulated in the Charter (IGU-CGE 2016), decisions around assessment in geography should be underpinned by theoretical and applied research in order to refine curriculum, pedagogic, and assessment practices used in the teaching and learning of geography. Cooperation and collaboration of researchers across the globe will move this agenda forward. This volume is a starting point.

References

Anderson LW, Krathwohl DR (2001) A taxonomy for learning, teaching, and assessing: a revision of Bloom's taxonomy of educational objectives. Longman, New York

Andrews KE, Tressler KD, Mintzes JJ (2008) Assessing environmental understanding: an application of the concept mapping strategy. Environ Educ Res 14(5):519–536

Anthamatten P (2010) Spatial thinking concepts in early grade-level geography standards. J Geogr 109(5):169–180

Atasoy E (2009) Eighth grade students' associations of geographical and environmental concepts. New Educ Rev 17(1):102–124

Baird J, Caro DH, Hopfenbeck TN (2016) Student perceptions of predictability of examination requirements and relationship with outcomes in high-stakes tests in Ireland. Ir Educ Stud 35(4):361–379

Baker DP, Wiseman AW (eds) (2009) Gender, equality and education from international and comparative perspectives, vol 10. Emerald Group Publishing, Bingley

Baldwin R (2003) Teachers' use of feedback comments to improve student achievement in standards-based assessment. Int J Learn 10:1197–1208

Battersby SE, Golledge RG, Marsh MJ (2006) Incidental learning of geospatial concepts across grade levels: map overlay. J Geogr 105(4):139–146

Bijsterbosch E (2018) The impact of national examinations on geography teachers' assessment practices in the Netherlands. Geogr Educ 31:41–48

Bijsterbosch E, van der Schee J, Kuiper W (2017) Meaningful learning and summative assessment in geography education: an analysis in secondary education in the Netherlands. Int Res Geogr Environ Educ 26(1):17–35. https://doi.org/10.1080/10382046.2016.1217076

Bijsterbosch E, van der Schee J, Kuiper W, Beneker T (2016) Geography teachers' practices regarding summative assessment: a study of pre-vocational education in the Netherlands. Rev Int Geogr Educ Online 6(2):118–134

Black P (2006) Assessment for learning: where is it now? Where is it going? In: Rust C (ed) Improving student learning through assessment. Oxford Centre for Staff and Learning Development, Oxford, pp 9–20

Black P, Wiliam D (2012) The reliability of assessments. In: Gardener J (ed) Assessment and learning, 2nd edn. Sage Publications, London, pp 198–213

Black P, Wiliam D (2018) Classroom assessment and pedagogy. Assess Educ Princ Pol Pract 25(6):551–575. https://doi.org/10.1080/0969594X.2018.1441807

Bourke T, Lane R (2017) The inclusion of geography in TIMSS: can consensus be reached? Int Res Geogr Environ Educ 26(2):166–176

Bourke T, Ryan M, Lloyd M (2016) The discursive positioning of graduating teachers in accreditation of teacher education programs. Teach Teach Educ 53:1–9

Brown G, Micklewright J, Schnepf V, Waldmann R (2005) International surveys of educational achievement: how robust are the findings? J Roy Stat Soc 170(3):623–646

Budke A, Schiefele U, Uhlenwinkel A (2010) 'I think it's stupid' is no argument: investigating how students argue in writing. Teach Geogr 35(2):66–69

Butt G, Weeden P, Chubb S, Srokoszc A (2006) The state of geography education in English secondary schools: an insight into practice and performance in assessment. Int Res Geogr Environ Educ 15(2):134–148

Cai J, Lester F (2007) Contributions from cross-national comparative studies to the internationalisation of mathematics education: studies of Chinese and US classrooms. In: Atweh B, Barton AC, Borba MC, Gough N, Keitel C, Vistro-Yu C, Vithal R (eds) Internationalisation and globalisation in mathematics and science education. Springer, Dordrecht, pp 269–283

Carless D (2007) Learning-Orientated Assessment: Conceptual Bases and Practical Implications Innovat Educ Teach Int 44(1):57–66

Cooper B, Cowie B (2010) Collaborative research for assessment for learning. Teach Teach Educ 26(4):979–986

Corcoran T, Mosher F, Rogat A (2009) Learning progressions in science: an evidence-based approach to reform. Consortium for Policy Research in Education, Philadelphia

Cox M, Elen J, Steegen A (2018) A test to measure students' systems thinking abilities in geography. Eur J Geogr 9(1):105–120

Crisp V (2008) Exploring the nature of examiner thinking during the process of examination marking. Camb J Educ 38(2):247–264

Davies P (2002) Levels of attainment in geography. Assess Educ Princ Pol Pract 9(2):185–204

Davies P, Durbin C, Clarke J, Dale J (2004) Developing students' conceptions of quality in geography. Curric J 15(1):19–34

Earl AM (2003) Assessment as learning: using classroom assessment to maximize student learning. Corwin Press, Thousand Oaks

Edelson DC, Shavelson RJ, Wertheim JA (eds) (2013) A road map for 21st century geography education: assessment—recommendations and guidelines for assessment in geography education (a report from the Assessment Committee of the Road Map for 21st Century Geography Education Project). National Geographic Society, Washington

Eze E (2021) Why secondary school geography students perform poorly in external examinations. J Geogr 120(2):51–60

Fjellborg AA, Kramming K (2021) Sustainable development: exploring gender differences in the Swedish national test in geography for Grade 9. Int Res Geogr Environ Educ. https://doi.org/10.1080/10382046.2021.1927366

Gibbs G, Simpson C (2004) Conditions under which assessment supports student learning. Learn Teach High Educ 1:3–31

Gipps C, Stobart G (2009) Fairness in assessment. In: Wyatt-Smith C, Cumming J (eds) Educational assessment in the 21st century: connecting theory and practice. Springer, Dordrecht, pp 105–118

Graham L, Tancredi H, Willis J, McGraw K (2018) Designing out barriers to student access and participation in secondary school assessment. Aust Educ Res 45(1):103–124. https://doi.org/10.1007/s13384-018-0266-y

Harlen W, James M (1997) Assessment and learning: differences and relationships between formative and summative assessment. Assess Educ Princ Pol Pract 4(3):365–379. https://doi.org/10.1080/0969594970040304

Harwood D, Rawlings K (2001) Assessing young children's freehand sketch maps of the world. Int Res Geogr Environ Educ 10(1):20–45

Huynh NT, Sharpe B (2013) An assessment instrument to measure geospatial thinking expertise. J Geogr 112(1):3–17

IGU-CGE (International Geographical Union Commission on Geographical Education) (2016) 2016 International Charter on Geographical Education. IGU-CGE, Beijing

Klenowski V (2014) Towards Fairer Assessment. Aust Educ Res 41(4):445–470

Koretz D (2009) How do American students measure up? Making sense of international comparisons. Future Child 19(1):37–51

Lane R, Bourke T (2017) Possibilities for an international assessment in geography. Int Res Geogr Environ Educ 26(1):71–85. https://doi.org/10.1080/10382046.2016.1165920
Lane R, Bourke T (2019) Assessment in geography education: a systematic review. Int Res Geogr Environ Educ 28(1):22–36. https://doi.org/10.1080/10382046.2017.1385348
Leat D, Nichols A (2000) Brains on the table: diagnostic and formative assessment through observation. Assess Educ Princ Pol Pract 7(1):103–121
Lee J, Bednarz R (2012) Components of spatial thinking: evidence from a spatial thinking ability test. J Geogr 111(1):15–26
Lock G, Barrett R (2002) Standards framework: developing scales of achievement in post-compulsory education: a case study. Issues Educ Res 12(1):35–48
Mansell W, James M, The Assessment Reform Group (2009) Assessment in schools. Fit for purpose? A commentary by the teaching and learning research programme. Economic and Social Research Council, London
McDowell L, Sambell K, Davison G (2009) Assessment for learning: a brief history and review of terminology. In: Rust C (ed) Improving student learning through the curriculum. Oxford Centre for Staff and Learning Development, Oxford, pp 56–64
Mishra RK (2015) Mapping the knowledge topography: a critical appraisal of geography textbook questions. Int Res Geogr Environ Educ 24(2):118–130
Munowenyu E (2007) Assessing the quality of essays using the SOLO taxonomy: effects of field and classroom-based experiences by 'A' level geography students. Int Res Geogr Environ Educ 16(1):21–43
National Research Council (2006) Learning to think spatially. The National Academies Press, Washington
Nursa'ban M, Kumaidi K, Mukminan M (2020) Factors of critical spatial thinking for a geography metacognition assessment in Indonesian senior high schools. Rev of Int Geog Educ 10(2):186–204
Panadero E, Tapia JA, Huertas JA (2012) Rubrics and self-assessment scripts effects on self-regulation, learning and self-efficacy in secondary education. Learn Indiv Differ 22(6):806–813
Pinar A (2011) Geography teachers' views on the assessment and evaluation instruments and methods used in the renewed geography curriculum. Educ Res Rev 6(3):334–341
Prais SJ (2007) Two recent (2003) international surveys of schooling attainments in mathematics: England's problems. Oxf Rev Educ 33(1):33–46
Riley K, Torrance H (2003) Big change question: as national policy-makers seek to find solutions to national education issues, do international comparisons such as TIMMS and PISA create a wider understanding, or do they serve to promote the orthodoxies of international agencies? J Educ Change 4:419–425
Sadler DR (1989) Formative assessment and the design of instructional systems. Instr Sci 18:119–144
Sadler DR (2007) Perils in the meticulous specification of goals and assessment criteria. Assess Educ 14(3):387–392
Sadler DR (2017) Academic achievement standards and quality assurance. Qual High Educ 23(2):81–99. https://doi.org/10.1080/13538322.2017.1356614
Scriven M (1967) The methodology of evaluation. In: Tyler RW, Gagné RM, Scriven M (eds) Perspectives of curriculum evaluation. Rand McNally, Chicago, pp 39–83
Solem M, Stoltman J, Lane R, Bourke T, Chang CH, Viehrig K (2018) An assessment framework and methodology for a Trends in International Geography Assessment Study (TIGAS). Geogr Educ 31:7–15
Stobart G (2008) Testing times: the uses and abuses of assessment. Routledge, Abingdon
Stoltman J, Lidstone J, Kidman G (2014) Geography and international assessment: opportunity or distraction. Int Res Geogr Environ Educ 23(3):193–196
Tiknaz Y, Sutton A (2006) Exploring the role of assessment tasks to promote formative assessment in Key Stage 3 Geography: evidence from twelve teachers. Assess Educ Princ Pol Pract 13(3):327–343

Tomaszewski B, Vodacek A, Parody R, Holt N (2015) Spatial thinking ability assessment in Rwandan secondary schools: baseline results. J Geogr 114(2):39–48
Torrance H (2007) Assessment as learning? How the use of explicit learning objectives, assessment criteria and feedback in post-secondary education and training can come to dominate learning. Assess Educ 14(3):281–294
van der Schee J, Kolkman R (2010) Multimedia tests and geographical education: the 2008 International Geography Olympiad. Int Res Geogr Environ Educ 19(4):283–293
Wehry S, Monroe-Ossi H, Cobb S, Fountain C (2012) Concept mapping strategies: content, tools and assessment for human geography. J Geogr 111(3):83–92
Wertheim JA, Edelson DC, The Road Map Project Assessment Committee (2013) A road map for improving geography assessment. Geogr Teach 10(1):15–21
Wiliam D (2018) Embedded formative assessment, 2nd edn. Solution Tree Press, Bloomington
Willis J, DeLuca C (2019) Quality assurance through assessment. In Peters M (ed) Encyclopedia of Teacher Education. Springer, Singapore
Wolf DP (1993) Assessment as an episode of learning. In: Bennett RE, Ward WC (eds) Construction versus choice in cognitive measurement: issues in constructed response, performance testing, and portfolio assessment. Lawrence Erlbaum Associates, Hillsdale, pp 213–240
Wrigley T (2004) School effectiveness: the problem of reductionism. Br Educ Res J 30(2):227–244
Yang D (2013) Comparing assessments within junior geography textbooks used in mainland China. J Geogr 112(2):58–67
Yasar O (2009) A comparative analysis of assessment and evaluation exercises included in geography textbooks written according to the 2005 secondary education geography curriculum and textbooks of the former curriculum in Turkey. Int J Prog Educ 5(1):45–68

Theresa Bourke is an Associate Professor at the Queensland University of Technology. She has held a number of leadership positions including Academic Program Director and Course Coordinator. She teaches into a number of curriculum and discipline units, specifically in geography. Her research interests include professional standards, professionalism, accreditation processes, assessment in geographical education, and teaching about, to, and for diversity.

Reece Mills commenced his career in education as a senior secondary school geography and science teacher. He is now a Senior Lecturer at the Queensland University of Technology, Brisbane. Reece teaches into undergraduate and postgraduate initial teacher education programs in the field of primary science. His research interests include science and sustainability education as well as primary science teacher education.

Part II
Assessment Related to Cross-Cutting Concepts and Skills in Geography

Chapter 2
Assessing Systems Thinking in Geography

Rainer Mehren and Armin Rempfler

Abstract On the grounds that geography is a system science, systems thinking or system competence is considered a fundamental concept of the school subject in the German Educational Standards in Geography. Accordingly, the Earth is understood as the biggest possible system consisting of a countless number of (geographical) complex, dynamic sub-systems. In contrast to non-systems thinking, a systems thinker takes superordinate principles of systems into account in the cognitive analysis and mental representation of geographical phenomena. This principle-led perspective provides deeper understanding of the internal and external interplay and complexity of systems, which may prevent human interference in such systems having unpredictable and unwanted adverse effects. In order to know what dimensions the construct 'systems thinking' consists of and which proficiency levels students can reach, there is a need for corresponding theoretical and empirical foundations and an assessment instrument for diagnosing, for example, whether certain interventions effectively promote systems thinking or not. This chapter primarily explains how a normative theoretically derived structure and stage model of geographic system competence was empirically validated. Finally, three item response models are evaluated, of which a two-dimensional model with the dimensions 'System organization and System behavior' and 'System-adequate intention to act' showed the best fit according to quality criteria. This provides a valid and reliable assessment tool for future diagnosing and promoting geographical system competence.

Disclaimer This chapter uses and builds on the previous work of Mehren et al. (2018) which is available from the *Journal of Research in Science Teaching*, Volume 55, Issue 5, pages 685–711 (© 2018 Wiley Periodicals, Inc.).

R. Mehren
University of Muenster, Münster, Germany
e-mail: rainer.mehren@uni-muenster.de

A. Rempfler (✉)
University of Teacher Education Lucerne, Lucerne, Switzerland
e-mail: armin.rempfler@phlu.ch

T. Bourke et al. (eds.), *Assessment in Geographical Education: An International Perspective*, Key Challenges in Geography,
https://doi.org/10.1007/978-3-030-95139-9_2

Keywords Systems thinking · Competence modelling · Socio-ecological approach · Item response theory

2.1 Introduction

Today's geography lessons deal with many of the great challenges of the twenty-first century, such as climate change, global migration, socio-economic disparities between North and South, ecological exploitation, or war for resources. These topics are highly complex and characterized by many influencing factors, such as intransparency, unpredictability, or instability. Due to these factors, there is a tendency for both teachers and students to oversimplify these complex problems, treating them in distorted ways and suggesting monocausal approaches as supposed solutions (cf. Booth-Sweeney and Sterman 2007). An alternative is to enable students to deal with complex situations by increasing their capacities to think at a systems level. Systems thinking in geography can be described as the ability to recognize, explain, and model the organization and behavior of complex natural and/or social environments at different scales as systems (Mehren et al. 2018). This way of thinking which goes beyond linear cause/effect relationships requires consideration of causes and their causes, or effects and their effects, and so on (Scheunpflug 2001), enabling systems forecasting and regulation based on the modelling.

With regard to geography education, it is imperative to consider (sub-)systems from both physical and human geography, not only in isolation, but also in their interaction as 'human–environment systems' (German Geographical Society 2014). Correspondingly, a theoretical foundation is required for an understanding of systems that do justice to the interrelationship between natural and social systems. The understanding of systems represented here (in detail in Mehren et al. 2018) is oriented towards 'social ecology'. This system understanding seeks to overcome the dichotomy between natural and social systems by viewing the interrelationship between society and nature as a system in its overall context (Hummel et al. 2017). The external relationships between society and nature therefore become internal relationships of the emerging socio-ecological system. Accordingly, the human–environment system approach is based on a more comprehensive system understanding than 'Earth System Science Education' (Orion and Libarkin 2014) which focuses purely on natural sciences.

2.1.1 Normative-Theoretical Structure and Stage Model

The starting point for the development of our theoretical model was superordinate principles of systems derived from social ecology (e.g., emergence, dynamics, regulation), which apply for both 'purely' natural and social systems as well as integrative human–environment systems (for more details see Mehren et al. 2018). These socio-ecological system principles were reconciled with the theoretical–conceptual designs and empirical findings already existing in the literature and a normative, initially four-dimensional competence model was developed (Table 2.1). Such a model structures a competence by dimensions (= structure model) and divides each of these dimensions into several proficiency levels (= stage model). The four dimensions derived from the literature were 'System organization', 'System behavior', 'System-adequate intention to act', and 'System-adequate action'.

The first two dimensions are 'System organization' and 'System behavior' (Sommer 2005; Sommer and Lücken 2010). Competence in reference to system organization (Dimension 1) means the ability to identify the organization of complex social and/or natural systems, and to model and describe their essential elements. System behavior (Dimension 2) encompasses the ability to analyze functions and behaviors of a system. Other authors present similar model concepts (Evagorou et al. 2009; Hmelo-Silver et al. 2007). The next two dimensions are 'System-adequate intention to act' and 'System-adequate action' (Köck and Rempfler 2004; Rost et al. 2003). While the focus is on the acquisition of knowledge for Dimensions 1 and 2, Dimensions 3 and 4 mainly involve the application of knowledge. Knowledge application occurs mentally or through action: mental application means system-adequate action in virtual space (Dimension 3). Actional application includes concrete action and therefore observable behavior in real space (Dimension 4). The ability for this type of comprehensive systems thinking and action corresponds to the definition of competence by Weinert (2001). For the empirical verification of our theoretical model, however, we limited ourselves to the 'System-adequate intention to act' (Dimension 3), since action in real space (Dimension 4) cannot be operationalized through large-scale assessment (Klieme and Leutner 2006).

The theoretical model development also included the division of Dimensions 1, 2, and 3 into further sub-dimensions ('System structure', 'System limit', 'System emergence', etc.) and three competence stages were distinguished for each sub-dimension (Table 2.1). Detailed descriptions as well as theoretical and, as far as possible, empirical justifications are available for all these sub-dimensions and stages. For reasons of space, these are not presented in detail here. They are described in a shortened version in Table 2.1 and further details including references can be found in Mehren et al. (2018, pp. 691–693). The stages relate to different levels of difficulty/complexity based on three difficulty-generating characteristics: number of elements and relations (low, moderate, high); type of networking (monocausal = A acts on B; linear = for example A acts on B, B acts on C; complex = for example A acts on B, B acts on C, and C in turn acts on A); and understanding of system-specific characteristics (such as non-linearity, emergence, limited predictability).

Table 2.1 Fundamental normative-theoretical model for geographic system competence

	Dimension 1: System organization Ability to identify a complex task environment in its organization as a system, and to model and describe its essential elements		**Dimension 2: System behavior** Ability to analyze functions and behaviors of a system			**Dimension 3: System-adequate intention to act** Ability of system-adequate action in mental space		**Dimension 4: System-adequate action** Ability of system-adequate action in real space
	System structure	**System limit**	**System emergence**	**System interaction**	**System dynamics**	**System prognosis**	**System regulation**	
Stage 1	• A small number of elements and relations is largely identified in an isolated manner • Low level of complexity • Monocausal thinking dominates	• Very vague differentiation of a set of relationships • Elements and relations are not viewed as part of a whole	• Focus on concrete, perceptible system components • Characteristics of the components are perceived as identical to the characteristics of the overall system	• Interaction is the result of proximity in time and space • Concept of causality has a monocausal focus • Simple stock and flow relationships are not identified to a significant extent	• The phenomenon or system is viewed as static-stable • Development trajectories are only considered from a monocausal perspective • Awareness of the temporal dimension is largely lacking	• Effects of system behavior are perceived vaguely and incidentally • Prognoses are based on direct and monocausal explanations • No awareness of limited predictability	• Regulative measures are implemented based on monocausal effect analysis • Effects of the intended measures are vaguely anticipated • Weak complexity reduction • System dynamics are not taken into account	

(continued)

Table 2.1 (continued)

	Dimension 1: System organization Ability to identify a complex task environment in its organization as a system, and to model and describe its essential elements		**Dimension 2: System behavior** Ability to analyze functions and behaviors of a system			**Dimension 3: System-adequate intention to act** Ability of system-adequate action in mental space		**Dimension 4: System-adequate action** Ability of system-adequate action in real space
	System structure	**System limit**	**System emergence**	**System interaction**	**System dynamics**	**System prognosis**	**System regulation**	
Stage 2	• A moderate number of elements and relations is increasingly identified together • Moderate level of complexity • Linear thinking dominates	• Moderate differentiation of a set of relationships • Integrated perspective is lacking, but elements and relations are no longer viewed exclusively in isolation	• Concrete, perceptible system components are combined at a higher level as part of a more general class with identical or similar characteristics	• Cause and effect are strictly separated • Interrelationships, series and parallel coupling are recognized • Simple stock and flow relationships are identified	• Longer linear relationships are also recognized • Causative explanation of movement and change • Developments are considered reversible	• Effects of system behavior are perceived systematically • Prognosis take a mono-final or multi-final direction • Vague awareness of limited predictability	• Regulative measures are implemented based on linear effect analysis • Effects of the intended measures are systematically anticipated • Moderate complexity reduction • System dynamics are sporadically taken into account	

(continued)

Table 2.1 (continued)

	Dimension 1: System organization Ability to identify a complex task environment in its organization as a system, and to model and describe its essential elements		**Dimension 2: System behavior** Ability to analyze functions and behaviors of a system			**Dimension 3: System-adequate intention to act** Ability of system-adequate action in mental space		**Dimension 4: System-adequate action** Ability of system-adequate action in real space
	System structure	**System limit**	**System emergence**	**System interaction**	**System dynamics**	**System prognosis**	**System regulation**	
Stage 3	• A large number of elements and relations is identified comprehensively and networked • High level of complexity • System is viewed as part of nested systems	• Clear differentiation of a set of relationships • Integrated perspective • Difference between neighboring systems is recognized	• Understanding that the interrelationship of system components results in new structures with new characteristics at a higher level (= emergence)	• Strict separation between cause and effect is eliminated • Feedback and cycles are recognized • Differentiation between internal system and external interaction • Complex stock and flow relationships are identified	• Non-linear (exponential and logistical) developments are also considered • The fact that developments are irreversible is recognized	• Effects of system behavior are recognized as interrelationship structures and taken into account in the prognosis • Differentiated awareness of limited predictability	• Regulative measures are implemented based on complex effect analysis • Effects of the intended measures are anticipated and measures are modified if applicable • Pronounced complexity reduction • System dynamics are continuously taken into account	

Source Rempfler and Uphues (2012, slightly modified)

2.2 Empirical Validation of the Model

Before the model in Table 2.1 could be empirically validated, it had to be simplified. Theoretically derived dimensions/sub-dimensions imply that they can be empirically separated from each other in order to diagnose and promote them independently from each other in learners. Since selective item development for the detailed definition of the sub-dimensions in Table 2.1 would not be feasible, they were compressed into the three dimensions and their core contents were abstracted to the dimension level. This means that the starting point of the validation process was a simpler model than the one in Table 2.1. The corresponding formulation of the dimensions and steps in this simplified model can be approximately reconstructed from Table 2.5, which is the empirically tested version (original three-dimensional version see Rempfler and Uphues 2012, p. 14). It is important to understand that the methodology and design of the empirical part of the study was aimed at developing test problems corresponding to the matrix structure of this simplified theoretical model. Subsequently, the model quality and fit were empirically tested. The actual empirical part consisted of three phases: problem development, pilot testing, and the main study (Fig. 2.1). The sections that follow will now elaborate on these three phases.

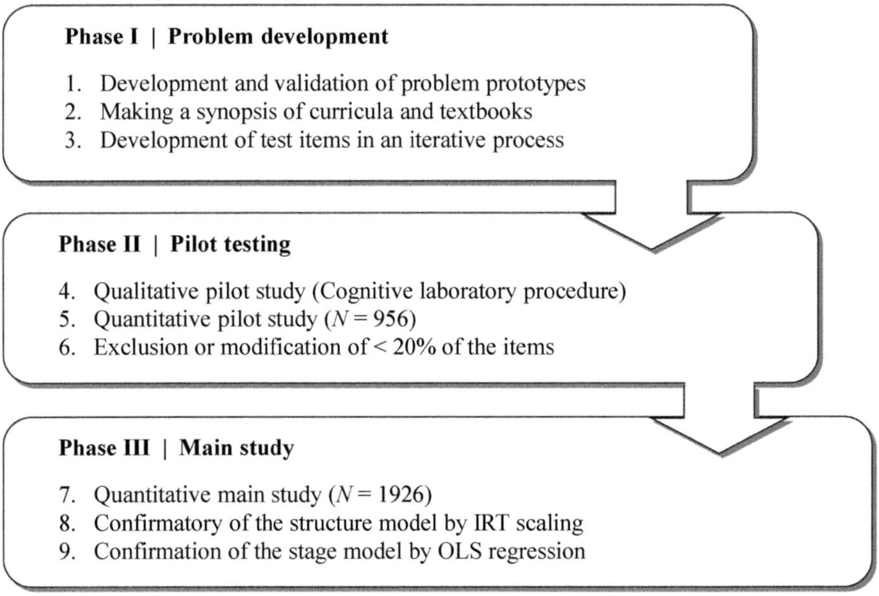

Fig. 2.1 Research design for the empirical validation of the competence model

Table 2.2 The topics of the 17 developed test problems

Physical geography systems	Plate tectonics in the Upper Rhine rift valley, global warming, trade wind cycle[a], land/sea breeze, water cycle (and salting of soil)
Human geography systems	Population in Africa, fair trade (banana), structural change in Munich, globalization of travel, migration, global textile trade, the US city
Human–environment systems	Nutrient cycle in the tropical rainforest, earthquake in Japan, rain in the city, Vesuvius, artificial lakes in the Alps

[a]see Fig. 2.2

Source Mehren et al., System competence modelling: Theoretical foundation and empirical validation of a model involving natural, social and human-environment systems, *Journal of Research in Science Teaching*, Wiley, © 2018 Wiley Periodicals, Inc.

2.2.1 Problem Development

During the first step of problem development, we created test items/problems prototypes, paying attention to the system construct, psychometric properties, and correctness in terms of content. After extensive revisions and validation in an iterative process, we generated 17 test problems for the main study with different geography content: approximately one third relating to physical geography systems, one third to human geography systems, and one third to human–environment systems. Details of this can be seen in Table 2.2.

The final questions comprise a master problem (see Fig. 2.2) and eight to ten related items (see Figs. 2.3, 2.4, and 2.5). A master problem was conceived so that it provides all relevant information for solving the items. In item development we ensured that the items do not build on each other, meaning they can be solved independently. Based on PISA (Programme for International Student Assessment), we chose the following answer formats in a 3:2:1 ratio: multiple choice, short answers, and extended responses (see Figs. 2.3, 2.4, and 2.5). We were also interested in concept maps as a special extended response problem type because they are well suited as a diagnosis tool for systems thinking (e.g., Nesbit and Adesope 2006).

Test items were generated for the three dimensions in the systems thinking model. During development, these were assigned to the respective competence dimensions and proficiency levels (Table 2.1). This a priori assignment to stages was performed based on the three difficulty-generating characteristics described above (Sect. 2.1.1).

2.2.2 Pilot Testing

Following Alavi (2005), Cohen (2000), and Long and Bourg (1996), test problems/items were piloted with two students, using a cognitive walk-through and a cognitive lab with paraphrasing and thinking aloud protocols. The results led to

Trade wind cycle

Twice a year in the spring (21 March, see Figure Aa) and autumn (23 September, see Figure Ac), the sun is vertically over the Equator.

During summer in the Northern Hemisphere, the vertical position of the sun shifts to the Tropic of Cancer (Figure Ab). The situation is reversed in winter: on 21 December, the sun's rays fall vertically onto the Tropic of Capricorn (Figure Ad).

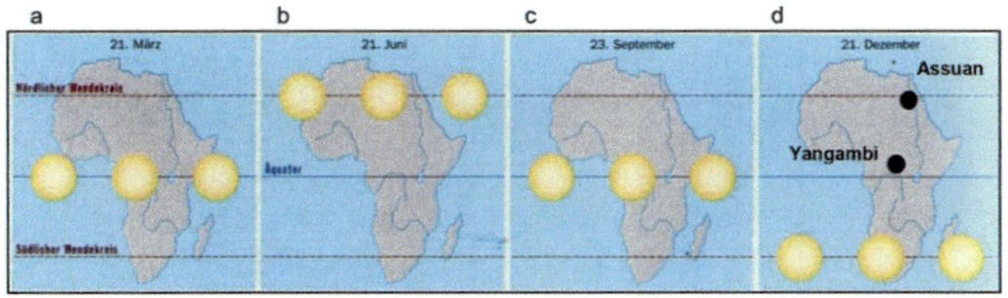

Figure A. Change in altitude of the sun

The sun incidence angle is related to the angle of the Earth's axis (Figure B). Where the sun's rays strike vertically, their effect is the greatest: the Earth's surface is heated intensely. The flatter the sun incidence angle, the larger the area over which the same radiation is distributed. In this case, the Earth's surface is not warmed nearly as much.

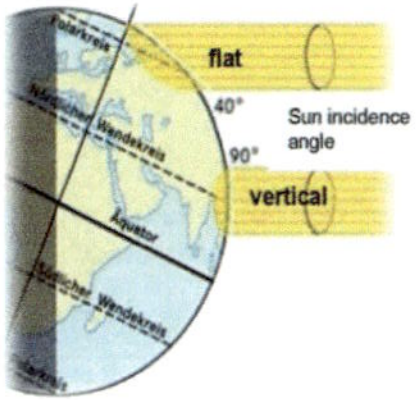

Figure B. Sun incidence angle on 21 June

When the sun is vertically over the Equator, high temperatures result close to the Earth's surface. This results in the evaporation of large volumes of water. The water which evaporates is carried upwards by air warmed close to the ground. Clouds form at higher altitudes, descending as rain on the same day (see thick arrows in Figure C). Air close to the ground rises because it expands as it is heated, which makes it lighter (lower density). This rising air creates a low-pressure area (L) close to the ground, and a high-pressure area (H) at higher altitudes. The air at higher altitudes cools down rapidly and wants to fall again (higher density), but is forced outwards by rising air. Upper winds develop as a result. The air can only drop gradually, reaching the ground at approximately 30° north and south. High-pressure areas form here. From here, the air flows back to the low-pressure area at the Equator as surface wind.

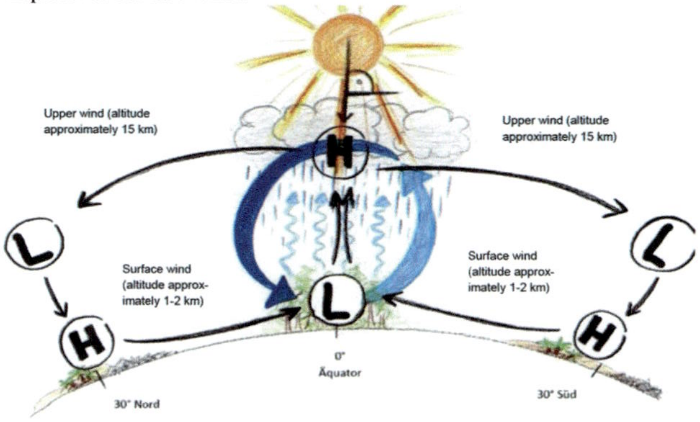

Figure C. Trade wind cycle

Fig. 2.2 Master problem for the topic 'Trade wind cycle'

Item 1

The relationship between two concepts can be presented in the form of an arrow. The arrow is labelled to make the relationship clear. Here is an example:

Vertical position of the sun —causes→ high temperatures

With the help of the text and illustrations in the master problem, express a similar relationship.

.... —...→ ....

Fig. 2.3 Item example for the answer format 'short answers'. The item represents the empirically validated Dimension 1 (System organization and System behavior) and Stage 1 (see Table 2.5)

Item 4

Figure D illustrates a cycle (excerpt from Figure C). How would you explain it to a friend?

Make sure that the relationships are clear, and that it becomes evident why it is called a cycle.

Explanation of Figure D:

....
....
....
....

Figure D. Daily weather in the tropics.

Fig. 2.4 Item example for the answer format 'extended response'. The item represents the empirically validated Dimension 1 (System organization and System behavior) and Stages 1–3 (see Table 2.5): Stage 1 = monocausal relationship, Stage 2 = linear relationship, Stage 3 = correct explanation of the whole cycle

a final revision of the test problems/items. Finally, a quantitative pilot study was conducted using the existing item pool with $N = 956$ school students (43.6% female; average age 15.39 years/$SD = 3.29$; Grade 9; secondary and grammar schools in Bavaria/Germany). The items were marked (interrater reliability, $\kappa = 0.68$), the number of correct responses were noted, and were analyzed using item response theory and a one-dimensional Rasch model (IRT; see Embretson and Reise 2000). The results of this analysis indicated that only a small proportion of the items (<20%) had to be excluded or modified for the main study.

Item 8

Winds at various altitudes are used to travel a certain route in a hot-air balloon. Heating air inside the balloon with a burner causes the balloon to rise. As the air gradually cools, the balloon falls again. One often 'flies' at an altitude of 3 km or even lower. The record altitude is 21 km.

You want to launch your balloon from Assuan (northern Africa) in mid-March. The location is marked on the information sheet (master problem, Fig. 2.2). Your objective is to reach the southern tip of Africa as quickly as possible. What is your approach?

Please check off one statement only.

You will use …
A. the ground winds to the Equator, then the ground winds from the Equator heading south.
B. the ground winds to the Equator, then the upper winds from the Equator heading south.
C. the upper winds to the Equator, then the upper winds from the Equator heading south.
D. the upper winds to the Equator, then the ground winds from the Equator heading south.

Fig. 2.5 Item example for the answer format 'multiple choice'. The item represents the empirically validated Dimension 2 (System-adequate intention to act) and Stage 2 (see Table 2.5). Correct is answer B

2.2.3 *Main Study*

Seventeen test booklets were used in the main study (see Table 2.2). Each randomly assigned test booklet contained two test problems, with an even coverage of topics across all booklets. A standardized procedure was used to instruct the classes at the start of testing that included the collection of demographic data such as age, gender, and primary language. The test administration incorporated a number of control mechanisms relating to reading speed and comprehension and intelligence (Schneider et al. 2017; Wilhelm et al. 2014). The test duration was 90 minutes.

The population for the main study was Grade 9 students ($N = 1926$ from 28 schools and 91 classes) in Bavaria/Germany (47.4% female, 0.9% not specified); 51.8% of the students in the study attended the *Gymnasium* (secondary-level grammar school), 26% the *Realschule* (secondary-level school), and 19.7% the *Hauptschule* (secondary-level school) (remaining 2.5%: *Wirtschaftsschule* [secondary-level commercial/vocational school]). In Germany, primary school ends after 4th grade. Secondary-level schools are differentiated into *Hauptschule* (lowest level), *Realschule* (medium level), and *Gymnasium* (highest level). Students of the *Hauptschule* and *Realschule* graduate after Grade 9 (around age 15) and then enter a career. The *Gymnasium* ends with Grade 12 and is a prerequisite for attending university. The average age of our study participants was 15.21 years ($SD = 0.674$) and 78.8% of them indicated that they mainly speak German at home.

The 154 items were marked as follows: multiple choice items were correct or incorrect; constructed-response items were correct, partially correct, or incorrect; and the concept maps were evaluated by calculating the structure index (see Mehren et al. 2018). After statistical analyses incorporating item response theory and Rasch modeling, 141 items remained.

For the confirmatory analysis of the factor structure, three different item response models were estimated (Table 2.3) and compared on the basis of model quality criteria. The models reflect the theoretical assumptions regarding the assignment of items to dimensions and are therefore differentiated in the number and significance of the dimensions. A single factor for all items was assumed in the simplest model (one-dimensional model). It therefore represents a general factor of geographic system competence and no dimensions are differentiated. The theoretically assumed Dimensions 1 and 2 were combined in model 2, resulting in a two-dimensional model. This corresponds to the considerations of various authors, according to which the intention to act has to be differentiated from a knowledge of system organization and behavior (Funke 2003; Rost et al. 2003). Model 3 reflects the three theoretical dimensions of the competence model underlying the item construction (System organization, System behavior, System-adequate intention to act).

The IRT scaling leads to estimates of empirical personal ability and item difficulty levels on a common scale, with item difficulty levels of particular interest here. The corresponding scale ranges from about –3 to +3, with the value 0 describing the average personal ability. Negative difficulty level values indicate that the item was easier than the average personal ability, while positive values indicate difficult items. For two-stage items ('correct', 'incorrect'; e.g., multiple choice like Fig. 2.5), the estimate resulted in exactly one item level of difficulty. It indicates the personal ability required to solve this item with a 50% probability, which means obtaining the point value 1 ('correct'). For multi-stage items on the other hand, a level of difficulty is estimated for each point value to be achieved. An item with a three-stage assessment for example ('incorrect', 'partially correct', 'correct'; see Fig. 2.4) was evaluated by the researchers considering two levels of difficulty: one that refers to obtaining the partially correct solution, and one that indicates the difficulty of obtaining a completely correct solution.

The item levels of difficulty expected in the competence model were related to the empirical difficulty levels to verify the stage model. Predicting the item levels of difficulty resulting from this model was attempted in an OLS (Ordinary Least Squares) regression at the manifest level through the theoretically assumed graduation of the items (expected difficulty level formulated during problem development). The theoretically assumed graduation is a three-stage variable on an ordinal scale. A dummy encoding was therefore performed for the regression so that three regression coefficients b resulted, specifying the average level of difficulty for Stage 1, the difference between Stages 1 and 2, and the difference between Stages 1 and 3 (see Table 2.4). The determination coefficient R^2 is used as a criterion for evaluating the prediction quality. This coefficient indicates the proportion of the variances in the empirical item levels of difficulty that is explained by the theoretically assumed graduation, and the highest possible value is desirable. For further data analysis and results of intelligence and reading comprehension, see Mehren et al. (2018).

Table 2.3 Quality criterion and schematic representation of the three estimated item response models

Model	BIC	Schematic representation
1D	47,015.34	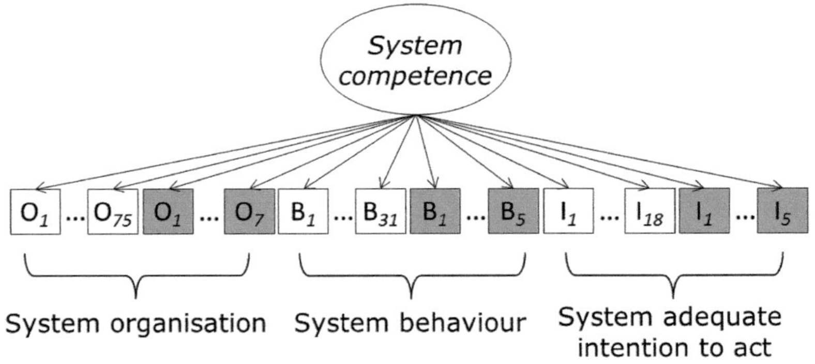
2D	47,010.68	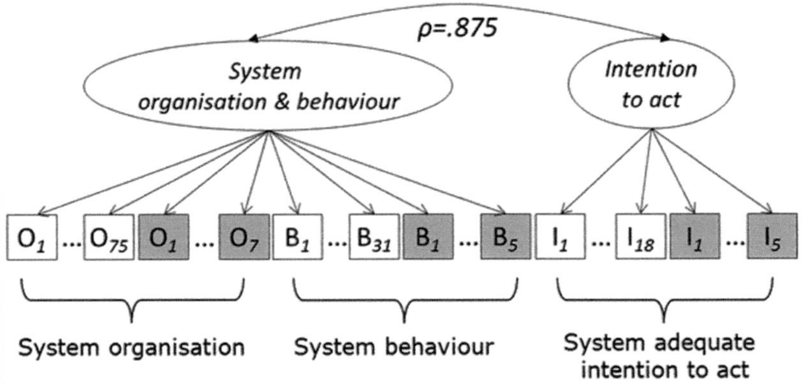

(continued)

Table 2.3 (continued)

Model	BIC	Schematic representation
3D	47,020.8	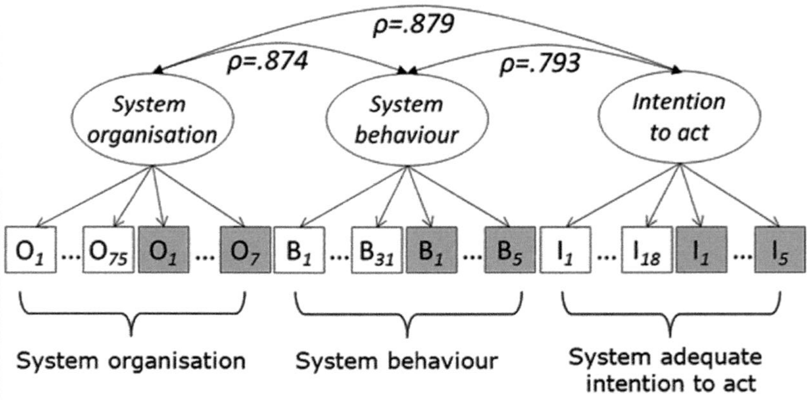

- BIC = Bayesian Information Criterion; the lower the BIC, the better the estimated model fits the empirical data
- ρ = latent correlation
- grey = concept map problems
- O = system organization problems, B = system behavior problems, I = system-adequate intention to act problems)

Source Mehren et al. (2018)

Table 2.4 Results of the OLS regression to predict the empirical item difficulty

Dimension	R^2		Regression coefficients		
			b	*t*	*p*
1	0.298	Difficulty Stage 1	−0.778	−5.591	<0.001
		Difficulty (Stage 1–Stage 2)	0.857	4.803	<0.001
		Difficulty (Stage 1–Stage 3)	1.791	8.484	<0.001
2	0.594	Difficulty Stage 1	−0.702	−2.720	0.010
		Difficulty (Stage 1–Stage 2)	0.700	2.306	0.027
		Difficulty (Stage 1–Stage 3)	2.284	6.850	<0.001

Source Mehren et al., System competence modelling: Theoretical foundation and empirical validation of a model involving natural, social and human-environment systems, *Journal of Research in Science Teaching*, Wiley, © 2018 Wiley Periodicals, Inc.

2.3 Core Results

2.3.1 *Empirically Validated Structure and Stage Model*

With respect to the BIC model quality criterion, the two-dimensional model has a better fit than the one-dimensional and three-dimensional models (see Table 2.3; further statistical characteristics, see Mehren et al. 2018). Based on the analyses of the model quality as well as further content-related considerations, a two-dimensional factor model of geographic system competence (Table 2.5) is therefore assumed, in which Dimension 1 (System organization and System behavior) is differentiated from Dimension 2 (System-adequate intention to act).

Figure 2.6 shows the scatter plots for the relationship between the theoretically assumed graduation and empirical item levels of difficulty for the items in the respective dimension. A positive relationship is discernible: the higher the assumed graduation of the item, the higher the item level of difficulty estimated by the model. Table 2.4 shows the results of the regression to predict the empirical item levels of difficulty on the basis of the assumed graduation. Accordingly, almost 30% of the variance in the difficulty of the items that measure the first dimension can be explained by the assumed graduation. For Dimension 2, the value is 59%. The results fit well with the theoretical assumption that a sufficient differentiation of personal abilities can be realized with three competence stages. The items and stages that should be more difficult according to the competence model proved to be more difficult according to the empirical trend, which is considered positive. Albeit the correlations are not perfect. This is because the final assignment of items to competence levels had in some cases to be corrected compared with the theoretically assumed graduation. However, one needs to consider that the theoretical graduation was based on a deductive approach, in which assumptions about the difficulty of the items were made before testing (a priori) by means of the difficulty-generating characteristics. Compared with inductive access, this approach has the advantage of defining thresholds between competence

Table 2.5 Theoretically founded and empirically validated two-dimensional model for geographic system competence

Competence stages	Competence dimensions	
	Dimension 1: **System organization and system behavior** **System structure, limit, emergence, interaction and dynamics** *Identifying a complex reality in the organization as a system and analyzing its functions and behaviors*	**Dimension 2:** **System-adequate intention to act** **System prognosis and regulation** *System adequate action in the mental space*
Stage 1	The student identifies a low number of elements and relations, mainly isolated or monocausal and as a vaguely differentiated set of relationships. His analysis of monocausal developments is based on a weakly developed functional and process understanding	The student develops prognoses and regulative measures based on the monocausal analysis of effects, vague anticipation of effects and weakly defined reduction of complexity
Stage 2	The student identifies a moderate number of elements and relations, mainly linear and as a moderately differentiated set of relationships. His analyses of linear developments is based on an understanding of interrelationships, series and parallel coupling as well as simple stock and flow relationships	The student develops prognoses and regulative measures based on the linear analysis of effects, anticipation of effects and moderately defined reduction of complexity
Stage 3	The student identifies a high number of elements and relations, mainly complex and as a highly differentiated set of relationships, and as part of nested systems. His analyses of linear and non-linear developments is based on an understanding of feedback and cycles as well as demanding stock and flow relationships, irreversibility and emergence	The student develops prognoses and regulative measures based on the complex analysis of effects, the anticipation of effects and highly defined reduction of complexity as well as awareness of limited predictability

Source Mehren et al., System competence modelling: Theoretical foundation and empirical validation of a model involving natural, social and human-environment systems, *Journal of Research in Science Teaching*, Wiley, © 2018 Wiley Periodicals, Inc.

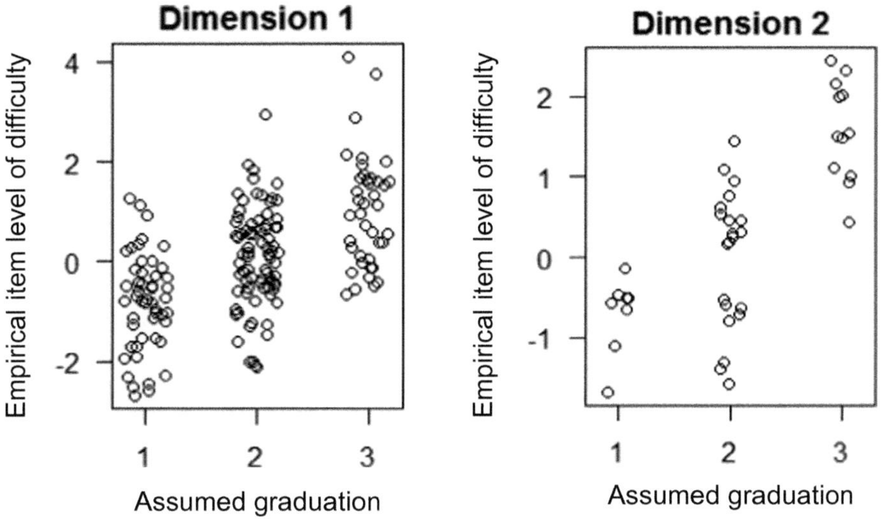

Fig. 2.6 Relationship between the assumed graduation and empirical item level of difficulty for all items in Dimension 1 (System organization and System behavior; left) and Dimension 2 (System-adequate intention to act; right) (*Source* Mehren et al., System competence modelling: Theoretical foundation and empirical validation of a model involving natural, social and human-environment systems, *Journal of Research in Science Teaching*, Wiley, © 2018 Wiley Periodicals, Inc.)

levels less arbitrarily, not according to numerical values but based on content and argumentation (see Seifert 2015).

2.3.2 Explanations Based on the 'Trade Wind Cycle' Problem/Test Items

Table 2.6 shows the relationship between assumed graduation and empirical item level of difficulty for all items of the 'Trade wind cycle' problem (Fig. 2.2). The cloud of points (in the upper left of Table 2.6) corresponds to a subset of the cases from the diagrams in Fig. 2.6, but not subdivided by dimensions. Again, there is a positive correlation, i.e., those items and levels that should be more difficult according to the competence model are more difficult in the trend. A classification of Table 2.6 in Fig. 2.6 also makes it clear that the items for the 'Trade wind cycle' problem are on the whole rather easy to moderately difficult compared with the other test problems. In particular, this applies to Dimension 1.

Detailed information on the coding and empirical difficulty of the items presented above (Figs. 2.3, 2.4, and 2.5) can also be found in Table 2.6. The coding shows how the corresponding item was evaluated. The percentage in parentheses refers to the relative number of students who solved the item incorrectly or correctly. Indirectly, it also gives an indication of the difficulty of the respective item. The corresponding

Table 2.6 Relationship between the assumed graduation and empirical item difficulty in the 'Trade wind cycle' problem

<table>
<tr><td colspan="2" rowspan="3">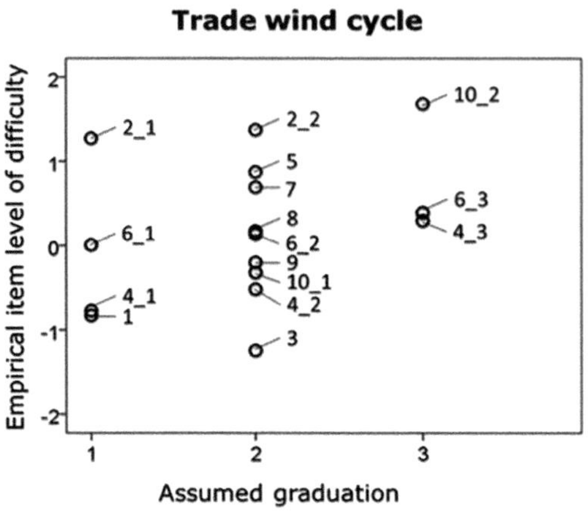
</td><td colspan="2">Item 4
(Partial item 4.1, 4.2, 4.3)</td></tr>
<tr><td>Coding</td><td>Graduated selection:
• Incorrect → Stage 0 (22.5%)
• Monocausal → Stage 1 (7.9%)
• Linear → Stage 2 (29.5%)
• Whole cycle → Stage 3 (40.1%)</td></tr>
<tr><td>Empirical item difficulty</td><td>• Stage 1 (= 4.1) = – 0.77 (one of the easier partial items of Stage 1)
• Stage 2 (= 4.2) = – 0.51 (one of the easier partial items of Stage 2)
• Stage 3 (= 4.3) = + 0.29 (easiest Stage 3 partial item)</td></tr>
<tr><td colspan="2">Item 1</td><td colspan="2">Item 8</td></tr>
<tr><td>Coding</td><td>Single selection:
• Incorrect → Stage 0 (31.9%)
• Correct → Stage 1 (68.1%)</td><td>Coding</td><td>Single selection:
• Incorrect → Stage 0 (53.8%)
• Correct → Stage 1 (46.2%)</td></tr>
<tr><td>Empirical item difficulty</td><td>Stage 1 = – 0.83 (easiest Stage 1 item of this test problem)</td><td>Empirical item difficulty</td><td>Stage 2 = + 0.17 (medium difficult item of Stage 2)</td></tr>
</table>

value is given in absolute terms; it can also be read directly from the graph (top left in Table 2.6). As theoretically expected, during the item development, this empirical data proved that monocausal tasks are rather easy (Item 1). Extended response items, which for example involve the recognition of a cycle, are generally more demanding (Item 4). Possibly, however, only a monocausal (Partial item 4.1) or linear relationship (Partial item 4.2) within the cycle is recognized, which is still rather easy. In the present case, apparently the recognition of the whole cycle (Partial item 4.3) was not so demanding either. We explain this by the fact that the graph used in Item 4 (Fig. 2.4) occurs identically—integrated into the overall graph—in the master problem (Fig. 2.2) and is carefully explained. In our view, the medium difficulty of Item 8 lies in the fact that items of Dimension 2 tend to cause more trouble because they involve the application of knowledge.

2.4 Promoting Systems Thinking—What Do We Already Know About It?

With the presented competence model, a theoretically founded measuring instrument is available, which allows, among other things, examination of whether certain teaching interventions promote systems thinking or not. This makes sense because the empirical evidence on questions like this is still relatively thin. Nevertheless, we would like to summarize what we consider to be the most important educational and methodological guiding principles that have so far been promising for the promotion of system competence in geography from a theoretical and empirical perspective:

1. *Promotion of concrete geographic content*: Systems thinking should not be promoted by means of theoretical-abstract system training, but rather by means of a correspondingly system-oriented preparation of geographic content (Riess et al. 2015). It must be shown how the individual aspects of a series of lessons are interlinked. System competence is generally regarded as domain-specific, which means that a good systems thinker in geography does not necessarily have to be a good systems thinker in physics.
2. *Involvement of both competence dimensions*: The competence dimensions validated in Table 2.5 are to be regarded as independent. For example, a student who can model a complex task environment as a system in a meaningful way is not necessarily capable of making recommendations for system-adequate action based on this model (and vice versa). Therefore, the promotion of both dimensions should be part of teaching (e.g., Mehren 2015). Dimension 2 (System-adequate intention to act) is often neglected, in that the systemic interrelationships are represented, but not prognostically or regulatively worked on (e.g., with 'What if … -questions', which are to be answered based on these interrelationships).
3. *Embedding in a cumulative approach*: We do not understand systems thinking as the addition of new teaching topics, but rather as an integrated teaching principle. Its potential can only develop realistically if it is not limited to a few isolated lessons but is seen as a guiding spiral-curricular design feature (Ossimitz 2000). There is empirical evidence that the ability to think systemically can also be lost (Bell 2003). The promotion of systems thinking is not reserved for higher grades but can be started in elementary school (Booth-Sweeney and Sterman 2007).
4. *Integration of graphical representations*: If a system is represented graphically, this contributes sustainably to the improvement of the systems thinking ability (Hildebrandt 2007). For example, students learn different steps of system analysis using concept mapping. Compared with other forms of representation, related information is immediately clear and strongly networked elements are visually more prominent (Nesbit and Adesope 2006, 2013). If learners create a concept map themselves, on average, significantly higher learning effects are shown than if they evaluate a given concept map (Schroeder et al. 2018). With concept maps, both dimensions of the model presented in Table 2.5 can

be considered, system modelling as well as system-adequate prognosis and regulation (Mehren et al. 2018).

5. *Computer simulations*: Complex systems are often difficult to convey in concrete terms, especially if the goal is to understand (long-term) dynamics and developments. Dynamic-interactive computer simulations offer an effective tool for conveying dynamic system properties, even if only rudimentary information on the effects is available (Hmelo-Silver and Azevedo 2006; Riess and Mischo 2010).
6. *Use of a task culture with pre-structured openness*: In the development of system competence, teaching arrangements that initiate intensive discussions and the negotiation of system modelling and system-adequate action strategies in an open but pre-structured setting are beneficial. Karkdijk and colleagues (2019) and Applis (2014), for example, demonstrate the effectiveness of the mystery method in dealing with complex systems. The results show, among other things, that this cooperative learning method increases the understanding and interpretation of systemic connections, and the ability to reflect and to deal with uncertainties.
7. *Metacognition*: The importance of such a fundamental concept as systems thinking lies not in its mentioning, but in its reflection with regard to deeper insights. This enables students to develop an understanding of geographical phenomena outside of school. The 'habits' presented by Waters Foundation (2010) are metacognitive thinking patterns (e.g., 'seeks to understand the big picture') that students apply when dealing with complex systems (Meister 2020). If such habits are repeatedly related to concrete geographical content in class and used as a stimulus for reflection, they support the students' process of understanding (Riess and Mischo 2010).
8. *Work with counterintuitive case studies*: Complex systems often behave counterintuitively. Unreflected naive theories (such as 'child labor is generally bad') influence the reception of complex situations and may constitute a learning barrier (Ben-Zvi-Assaraf and Orion 2005). The comprehensive treatment of such case studies (e.g., Mehren 2015), which often generate hasty solutions in the minds of students, makes them aware of the necessity of a system approach and makes learning progression transparent.
9. *System-oriented conception of teaching*: In the initiation of systems thinking, the teacher is empirically shown to be a central factor (Hmelo-Silver and Azevedo 2006). He/she has to prepare content knowledge in a system-oriented manner in order to initiate a system understanding among students (Jacobson and Wilensky 2006). Teachers can only achieve this if they themselves have systemically understood the complexity of the topic. However, a conceptual understanding of teaching—e.g., in terms of systems thinking—is currently only poorly developed among teachers (Skaza and Carroll 2013).
10. *Allowing complexity*: Instruction is often based on the textbook double-page principle. Individual aspects of a topic are processed in small, self-contained sub-systems without linking them to previous or subsequent content. This leads to additive thinking by students who, at the end of a lesson, are for example

not able to explain how the short, closed nutrient cycling, shifting cultivation, palm oil plantation, biodiversity, and fair trade in the tropics are related. In this respect, it is important to continuously combine the partial aspects in order to develop an understanding of the overall system with its sub-systems (Mehren et al. 2014).

11. *Bearing uncertainty:* Complex problems that cause experts worldwide to fail can hardly be solved by students in 45 or 50 minutes. Here it is important to strengthen the students' (and teachers') tolerance of ambiguity and the ability to bear uncertainty (Furnham and Ribchester 1995). Students gain confidence not by pretending to find a false solution, but by looking at a problem from different sides and realizing that one has profoundly understood it but may not be able to solve it (Applis 2012).

2.5 Outlook

The competence model empirically confirmed in Table 2.5 is primarily oriented towards quantitative system characteristics. However, from a systems theory perspective, qualitative characteristics are more important for a profound understanding of complex dynamic systems. Only the third of the three difficulty-generating characteristics we used for the stage formation is qualitatively oriented (see Sect. 2.1.1). This characteristic shows weaknesses, in that the system-specific properties in the model (Table 2.5) are partly proven imprecisely (e.g., Stage 1 'weakly developed functional and process understanding'). The system properties evident at the higher levels, such as series and parallel coupling, emergence, etc., serve primarily to illustrate the model. As mentioned above, we had initially developed a detailed model (Table 2.1), but due to the particularly weak empirical state of knowledge at that time, we did not consider a selective item development for the detailed definition of the sub-dimensions to be possible. Accordingly, the existing model does not allow us to assess the students' competences in terms of concrete system-specific characteristics such as non-linearity, irreversibility, etc. This approach is in line with that of most international studies. Only very few have differentiated certain system-specific properties in test items. However, whether these task characteristics also correspond to differentiable personal characteristics, or whether the different tasks measure de facto the same, remains a central research priority. In this respect, the more recent scientific studies on 'tipping points' represent a fundamental expansion of knowledge in the understanding of functions and processes in natural, social, and human–environment systems (e.g., Lenton et al. 2008, 2019; Milkoreit et al. 2018). The analysis of these findings now forms the basis for us to derive theoretically and empirically founded system-specific characteristics that can be really operationalized and thus a qualitative understanding of systems can be acquired by learners (Fögele et al. 2020). This would eliminate the deficit described above, i.e., with the extended model and corresponding test items, we would be able to measure students' ability to recognize system properties such as 'non-linear dynamics', 'feedback', or 'limited predictability' with high selectivity.

References

Alavi SM (2005) On the adequacy of verbal protocols in examining an underlying construct of a test. Stud Educ Eval 31:1–26

Applis S (2012) Wertorientierter Geographieunterricht im Kontext Globales Lernen. Theoretische Fundierung und empirische Untersuchung mit Hilfe der dokumentarischen Methode. Geographiedidaktische Forschungen, 51. HGD, Weingarten

Applis S (2014) Global learning in a geography course using the mystery method as an approach to complex issues. Rev Int Geogr Educ Online 4(1):58–70

Bell T (2003) Strukturprinzipien der Selbstregulation: Studien zum Physiklernen, 25. Logos, Berlin

Ben-Zvi-Assaraf O, Orion N (2005) Development of system thinking skills in the context of earth system education. J Res Sci Teach 42(5):518–560

Booth-Sweeney L, Sterman JD (2007) Thinking about systems: student and teacher conceptions of natural and social systems. Syst Dynam Rev 23(2–3):285–312

Cohen AD (2000) Exploring strategies in test-taking. Fine-tuning verbal reports from respondents. In: Ekbatani G, Pierson H (eds) Learner-directed assessment in ESL. Erlbaum, Mahwah, pp 127–150

Embretson SE, Reise S (2000) Item response theory for psychologists. Erlbaum, Mahwah

Evagorou M, Korfiatis K, Nicolaou C, Constantinou C (2009) An investigation of the potential of interactive simulations for developing system thinking skills in elementary school: a case study with fifth-graders and sixth-graders. Int J Sci Educ 31(5):655–674

Fögele J, Mehren R, Rempfler A (2020) Tipping points – Schlüssel zum tiefgründigen Verständnis komplexer dynamischer Systeme bei Lernenden? Zeitschrift Für Geographiedidaktik/j Geogr Educ 48(3):83–100

Funke J (2003) Problemlösendes Denken. Kohlhammer, Stuttgart

Furnham A, Ribchester T (1995) Tolerance of ambiguity: a review of the concept, its measurement and applications. Curr Psychol 14:179–199

German Geographical Society (ed) (2014) Educational standards in geography for the intermediate school certificate with sample assignments. German Geographical Society, Bonn

Hildebrandt K (2007) Die Wirkung systemischer Darstellungsformen und multiperspektivischer Wissensrepräsentationen auf das Verständnis des globalen Kohlenstoffkreislaufs. Dissertation, Kiel

Hmelo-Silver CE, Azevedo R (2006) Understanding complex systems: some core challenges. J Learn Sci 15(1):53–61

Hmelo-Silver CE, Marathe S, Liu L (2007) Fish swim, rocks sit, and lungs breathe: expert-novice understanding of complex systems. J Learn Sci 16(3):307–331

Hummel D, Jahn T, Keil F, Liehr S, Stiess I (2017) Social ecology as critical, transdisciplinary science—conceptualizing, analyzing and shaping societal relations to nature. Sustainability 9(7):1050

Jacobson MJ, Wilensky U (2006) Complex systems in education: scientific and educational importance and implications for the learning sciences. J Learn Sci 15(1):11–34

Karkdijk J, Van der Schee JA, Admiraal WF (2019) Students' geographical relational thinking when solving mysteries. Int Res Geogr Environ Educ 28(1):5–21

Klieme E, Leutner D (2006) Kompetenzmodelle zur Erfassung individueller Lernergebnisse und zur Bilanzierung von Bildungsprozessen. Zeitschrift Für Pädagogik 52(6):876–903

Köck H, Rempfler A (2004) Erkenntnisleitende Ansätze – Schlüssel zur Profilierung des Geographieunterrichts. Aulis, Köln

Lenton TM, Held H, Kriegler E, Hall JW, Lucht W, Rahmstorf S, Schellnhuber HJ (2008) Tipping elements in the Earth's climate system. In: Proceedings of the National Academy of Sciences 105(6):1786–1793
Lenton TM, Rockström J, Gaffney O, Rahmstorf S, Richardson K, Steffen W, Schellnhuber HJ (2019) Climate tipping points—too risky to bet against. Nature 575(7784):592–595
Long DL, Bourg T (1996) Thinking-aloud: telling a story about a story. Discourse Process 21:329–339
Mehren M (2015) Piraterie vor Somalia. Ein komplexes Problem systemisch betrachtet. Geographie aktuell und Schule 37(215):30–40
Mehren R, Rempfler A, Buchholz J, Hartig J, Ulrich-Riedhammer EM (2018) System competence modelling: theoretical foundation and empirical validation of a model involving natural, social, and human-environment systems. J Res Sci Teach 55(5):685–711
Mehren R, Rempfler A, Ulrich-Riedhammer E (2014) Denken in komplexen Zusammenhängen. Praxis. Geographie 44(4):4–8
Meister J (2020) Eine videogestützte Prozess- und Produktanalyse der Systemkompetenz – am Beispiel der Bearbeitung eines Misterys. Dissertation. http://geb.uni-giessen.de/geb/volltexte/2020/15449. Accessed 9 Sept 2020
Milkoreit M, Hodbod J, Baggio J, Benessaiah K, Calderón-Contreras R, Donges JF, Mathias J-D, Rocha JC, Schoon M, Werners SE (2018) Defining tipping points for social-ecological systems scholarship—an interdisciplinary literature review. Environ Res Lett 13(3):1–12
Nesbit JC, Adesope OO (2006) Learning with concept and knowledge maps: a meta-analysis. Rev Educ Res 76:413–448
Nesbit JC, Adesope OO (2013) Concept maps for learning: theory, research, and design. In: Schraw G, McCrudden MT, Robinson D (eds) Current perspectives on cognition, learning, and instruction. Information Age Publishing, Charlotte, Learning through visual displays, pp 303–328
Orion N, Libarkin J (2014) Earth system science education. In: Lederman NG, Abell SK (eds) Handbook of research on science education, vol. II. Routledge, New York, pp 481–496
Ossimitz G (2000) Teaching system dynamics and systems thinking in Austria and Germany. In: Proceedings of the 18th International Conference of the System Dynamics Society, Bergen, Norway
Rempfler A, Uphues R (2012) System competence in geography education. Development of competence models, diagnosing pupils' achievement. Eur J Geogr 3(1):6–22
Riess W, Mischo C (2010) Promoting systems thinking through biology lessons. Int J Sci Educ 32(6):705–725
Riess W, Schuler S, Hörsch C (2015) Wie lässt sich systemisches Denken vermitteln und fördern? Theoretische Grundlagen und praktische Umsetzung am Beispiel eines Seminars für Lehramtsstudierende. Geographie Aktuell Und Schule 37(215):16–29
Rost J, Lauströer A, Raack N (2003) Kompetenzmodelle einer Bildung für Nachhaltigkeit. Praxis der Naturwissenschaften – Chemie 8(52):10–15
Scheunpflug A (2001) Biologische Grundlagen des Lernens. Cornelsen, Berlin
Schneider W, Schlagmüller M, Ennemoser M (2017) Lesegeschwindigkeits- und Verständnistest für die Klassen 5–12+. Hogrefe, Göttingen
Schroeder NL, Nesbit JC, Anguiano CJ, Adesope OO (2018) Studying and constructing concept maps: a meta-analysis. Educ Psychol Rev 30(2):431–455
Seifert A (2015) Kompetenzforschung in den Fachdidaktiken auf der Grundlage von IRT-Modellen. In: Riegel U, Schubert S, Siebert-Ott G, Macha K (eds) Kompetenzmodellierung und Kompetenzmessung in den Fachdidaktiken. Fachdidaktische Forschungen 7:131–161
Skaza H, Carroll K (2013) Teachers' barriers to introducing system dynamics in K–12 STEM curriculum. Syst Dynam Rev 29(3):157–169
Sommer C (2005) Untersuchung der Systemkompetenz von Grundschülern im Bereich Biologie. Dissertation. http://macau.uni-kiel.de/receive/dissertation_diss_00001652. Accessed 4 Sept 2020
Sommer C, Lücken M (2010) System competenceare elementary students able to deal with a biological system? Nord Stud Sci Educ 6(2):125–143

Waters Foundation (2010) Systems thinking in schools: a Waters Foundation Project. Waters Foundation, Pittsburg

Weinert FE (2001) Concept of competence: a conceptual clarification. In: Rychen D, Salganik LH (eds) Defining and selecting key competencies. Hogrefe, Seattle, pp 45–65

Wilhelm O, Schroeders U, Schipolowski S (2014) BEFKI 8–10. Berliner Test zur Erfassung fluider und kristalliner Intelligenz für die 8. bis 10. Jahrgangsstufe. Hogrefe, Göttingen

Rainer Mehren is a Professor of Geography Education at the University of Muenster (Germany). His work and research focuses on global learning, systems thinking, ethical judgment, spatial orientation, and transfer of knowledge.

Armin Rempfler is a Professor of Geography Education at the University of Teacher Education Lucerne (Switzerland). His work and research focuses on systems thinking, learning in out-of-school learning settings, and instructional materials development/evaluation.

Chapter 3
Assessment of Mapping Learning Progressions in Geography

Jinhee Lee and Injeong Jo

Abstract This chapter aims to (1) introduce the methods that are commonly used in learning progression (LP) research and (2) discuss how an assessment is applied to empirical LP research in geography education. The first step in identifying an LP is to define what students know and can do at different levels of their learning of a concept. This study selected six map elements—contour lines, distance, location, map projection, orientation, and scale—as essential concepts in map understanding based on an extensive review and analysis of prior studies. A hypothesized LP on map understanding was proposed, and a map understanding test using ordered multiple-choice items to measure students' LP was developed. The knowledge and skill descriptions presented in the hypothesized LP were transformed into choices for each question item. After the main test was conducted, the coded students' response data was analyzed by the partial credit Rasch model. The research methodology and the assessment tool introduced in this chapter will provide a foundation for developing assessment tools to identify LPs in other areas of geographic knowledge, skills, and practices.

Keywords Geography education · Learning progression · Map understanding · Assessment of learning · Ordered multiple-choice

---

This chapter is based on a dissertation, Learning progression on map understanding: A case study of elementary through college students in Korea by Jinhee Lee, submitted to Texas State University August 2017.

J. Lee (✉)
Pusan National University, Busan, South Korea
e-mail: jinhee@pusan.ac.kr

I. Jo
Texas State University, San Marcos, TX, USA
e-mail: ijo@txstate.edu

T. Bourke et al. (eds.), *Assessment in Geographical Education: An International Perspective*, Key Challenges in Geography,
https://doi.org/10.1007/978-3-030-95139-9_3

3.1 Introduction

A learning progression (LP) can be defined as 'empirically-grounded and testable hypotheses about how students' understanding of, and ability to use, core scientific concepts and explanations and related scientific practices grow and become more sophisticated over time, with appropriate instruction' (Corcoran et al. 2009, p. 15). Validating an LP helps obtain empirical data to confirm students' learning processes in line with disciplinary experts' and educators' suggestions (Duncan and Hmelo-Silver 2009). The final goal of LP research is to deliver evidence of students' learning and their degree of understanding within and between grade levels (Solem et al. 2014). This information can be utilized to set proper curriculum scope and sequences that reflect students' developmental stages (Corcoran et al. 2009; National Research Council 2007). Recently, US researchers in the field of science education actively conducted LP research. Their findings became key concepts in the development of "A Framework for K–12 Science Education" (Maeng et al. 2013) This framework 'guide[s] the development of new standards that in turn guide revisions to science-related curriculum, instruction, assessment, and professional development for educators' (National Research Council 2012, p. 2).

In geography education, LPs for essential geographic concepts are still emerging as a research area. The *Road Map for 21st Century Geography Education* was the first document that discussed the necessity of LP research in geography education (Bednarz et al. 2013). Among 13 recommendations for geography education research, the first recommendation stated, 'The committee recommends that geography education researchers engage in systematic efforts to identify learning progressions in geography both within and across grade bands (e.g., grades K–4, 5–8, 9–12)' (Bednarz et al. 2013, p. 8). LP research was viewed as helpful for setting up core geographic ideas associated with students' learning such as revising curriculum, assessments, and learning activities (Lee 2017). Despite this foundational progress in the field, this research is still in its infancy (Solem et al. 2014; Huynh et al. 2015). The reason for such paucity of research is due to a lack of understanding of how to develop and use assessments to measure students' LPs (Lee 2017). The most important step in empirical LP research is to develop an assessment tool in a reliable and valid way (Duncan and Hmelo-Silver 2009).

This chapter aims to take research in this field forward by (1) introducing an assessment that is commonly used in LP research, and (2) discussing how assessment is applied to empirical LP research in geography education. It is hoped that the content covered in this chapter will help readers understand how assessment is used and how important it is for LP research in geography education.

3.2 Research Approaches to Identify Learning Progressions

Three research approaches are commonly used to identify LPs. The first is a data-grounded approach (Huynh et al. 2015). This approach relies on the synthesis of academic literature, curriculum frameworks, and the researcher's/teacher's insights into statements that represent students' learning at different stages. For example, in relation to central place theory, possible ideas for 3rd and 4th grade students (aged 9–10 years) are 'a central place is created where it is crowded with people', while a possible idea for 7th and 8th grade students (aged 13–14 years) is 'the central place is a place that provides goods and services to the surrounding area'. However, there are limitations with this approach, in that further validation studies need to be conducted to guarantee reliability and validity (Lee and Liu 2010). Even if prior studies have reported that students go through a specific developmental process, it cannot be affirmed whether current students go through the same stages and develop their level of understanding. Therefore, it is necessary to verify that students' understandings of key concepts are the same as the results of prior studies. The most widely used mechanism for such verification is to develop an assessment and measure a student's LP through it.

The second approach is 'in situ' student understanding (Huynh et al. 2015). This approach not only collects data from institutional and academic literature, but validates the collected data through assessments, clinical interviews, and teaching episodes (Clements et al. 2004). An example of the in situ approach is Maeng and Lee's (2015) LP in understanding plate tectonics. These authors identified conceptual constructs of plate tectonics from previously published studies, developed an assessment tool to validate these constructs, and in doing so, identified an LP for plate tectonics. This second approach goes further than the first approach in that validation with students occurs.

The third approach is instructional interventions (Huynh et al. 2015) to identify the influence of the instruction on students' understanding and skills (Plummer and Krajcik 2010). For example, Songer et al. (2009) identified a LP on biodiversity for 4th–6th grade students (aged 10–12 years) using this approach. They examined previous studies and drew a hypothesized LP on biodiversity. Next, they developed instructional activities and assessments based on the hypothesized LP and presented them to 4th–6th grade students for an eight-week period. The researchers then refined the hypothesized LP after reflecting on students' test results.

The first approach, which is data-grounded, has the benefit of simplicity as it requires a conceptual analysis of institutional and academic literature only. However, as already mentioned, there are limitations in terms of validity and reliability (Huynh et al. 2015). The second approach has the advantage of requiring validity and reliability measures (Alonzo and Steedle 2009). This method can reliably derive the LPs of general students in specific contexts but does not account for variables such as learning/teaching strategies, instructors, and learning frequency (Maeng and Lee 2015). The third approach is beneficial in providing information about the nature and quality of explanations students develop for the hypothesized LP (Lee 2017).

In addition, more refined data about students' understanding on concepts as a result of instructional intervention can be identified over time (Songer et al. 2009). In this approach, it is important to develop and apply valid assessment tools that can accurately measure students' LP. To this end, various methods are used, including paper–pencil tests, observations, interviews, and evaluation through activities (Songer et al. 2009). Of the three approaches, the second approach, in situ student understanding, is the most common research method used in science education where most of the LP research is located (Maeng et al. 2013).

The purpose of this chapter is to show how to conduct LP research and how an assessment is used to measure students' LP using an exemplar study. We achieve this by showing how to empirically derive students' LP on map understanding, focusing on six map elements—contour lines, distance, location, map projection, orientation, and scale—as essential concepts in map understanding based on an extensive review and analysis of prior studies about students' development in map learning and spatial skills. First, we detail a process for developing an LP using the in situ approach and the assessment triangle framework. Second, a hypothesized LP on map understanding is then proposed. Third, the assessment tool using ordered multiple-choice items to measure students' LP is outlined alongside the analysis of assessment data from 1,486 Korean students. Finally, after analyzing the results of the assessment, the LP for map understanding is derived. The research procedure for this study is summarized in Fig. 3.1.

3.2.1 Learning Progressions Using the In Situ *Approach and Assessment Triangle Framework*

LP research using the in situ approach has three main steps: (1) to define learning performance indicators, (2) to devise assessment tools to measure students' level of understanding, and (3) to verify validity. An assessment is designed based on an hypothesized LP, empirical data that supports the hypothesis is collected, and items or tasks are used to test the hypothesis to see if components work (Mislevy et al. 2003). Assessment triangle is a widely used framework in LP research because it guides these three steps (Fig. 3.2) (National Research Council 2001; Maeng et al. 2013). This study also follows this framework to derive students' LP on map understanding.

The National Research Council (2006) introduced the assessment triangle as a useful way to develop an assessment tool in science education. Many recent LP studies have been conducted in accordance with this framework (e.g., Duschl et al. 2011; Johnson and Tymms 2011). The assessment triangle consists of three components (cognition, observation, and interpretation) that work recurrently (National Research Council 2001, 2006). *Cognition* is the first stage, where the researchers clearly set the hypothesized LP based on prior research about students' knowledge and skills, including how knowledge is acquired, and what knowledge and skills are important to evaluate students. During *Observation*, students' level of learning is

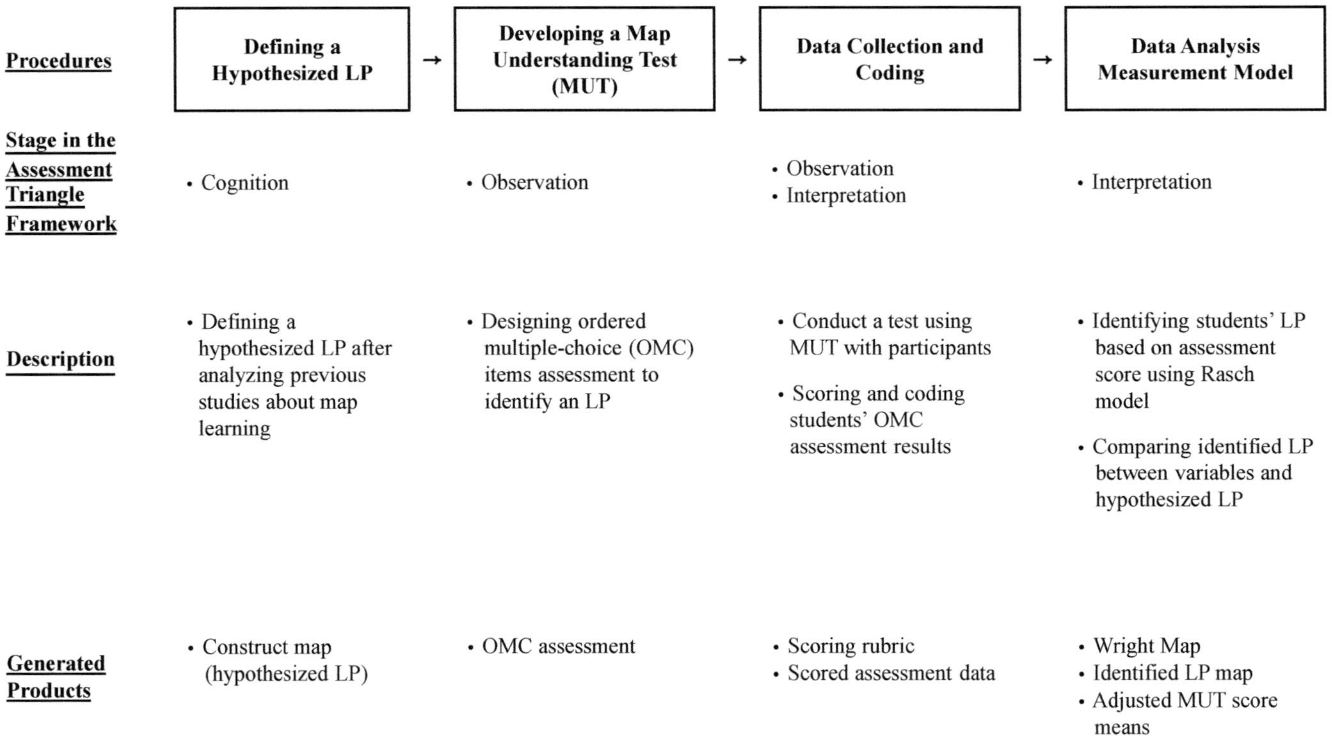

Fig. 3.1 Research procedure in this study (*modified in* Lee 2017, p. 47)

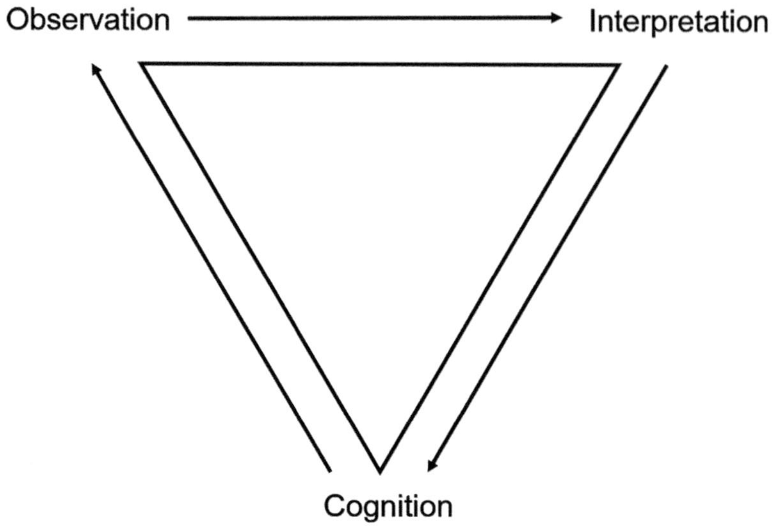

Fig. 3.2 The assessment triangle (*Source* National Research Council 2001, p. 112)

measured and evaluated by proper assessment tools. The tools must clearly measure students' levels and processes of understanding. Finally, *Interpretation* accounts for students' learning characteristics by interpreting the results of the assessments. Researchers can use quantitative or qualitative methods to determine patterns in the collected data. During the interpretation component, the hypothetical LP is enhanced into a verified LP underpinned by the empirical data. When all three components work properly, the tool that accurately measures students' LPs can be created (Pellegrino 2009). Iterative cycles of empirical testing and revision of assessment can derive more accurate LPs (Chang et al. 2020).

3.3 Defining a Hypothesized Learning Progression

In this study, map understanding was the chosen concept for the development of an LP. Map understanding consists of two parts. The first is comprehension of concepts and basic map elements including location, scale, distance, orientation, contour lines, and map projection. The second is understanding and using various types of thematic maps, such as a dot density map, a choropleth map, and a flow map. The present study focuses on the first part, understanding concepts and basic map elements, as the scope of study. This study was conducted through a synthesis of prior research. For example, Blades and Spencer (1986) designed a test for 4–6-year-old children's simple map reading ability. Boardman (1983, 1989) studied difficulties that students have with graphicacy and proposed instructional strategies that can be applied to solve the problems they have. These studies, however, are often criticized for their reliability because most of them were either conducted with a small number of students or were based on the researcher's personal experiences. In addition, studies conducted by other researchers on similar subjects were not combined but remained as separate studies. When individual studies are gathered and analyzed together, academically beneficial results that were not noticeable in individual analysis can now be seen. This approach is useful in research on LP (Bednarz et al. 2013).

Six map elements were chosen for inclusion in this LP after reviewing and analyzing 21 studies about students' learning about map concepts and spatial skills (see Table 3.1). These are: contour lines, distance, location, map projection, orientation, and scale. This study selected these six elements as essential concepts in map understanding (Lee 2017). Table 3.2 displays the hypothesized LP as it related to these six concepts, from the most basic (Level 1; 3rd grade, 8 years old) to sophisticated (Level 4; 12th grade, 17 years old).

3.4 Developing a Map Understanding Test: Ordered Multiple-Choice Items

The next step involved the development of an assessment to obtain evidence of what students know and can do at various levels identified on the hypothesized LP (Lee 2017). Ordered multiple-choice (OMC) items are widely used to identify and classify students' levels of understanding (Alonzo and Steedle 2009). While there is only one correct answer in traditional multiple choice questions, in aa OMC item, all choices are correct but represent different stages of development (Briggs and Alonzo 2012). Students choose the responses that best match their knowledge from the given alternatives. When developing OMC items, therefore, it is important to ensure that each choice represents a distinct level of competency. Level 1 choices correspond with Level 1 competency, Level 2 choices for students with Level 2 competency, and so forth. There are two reasons for using OMC questions to identify students' LP. First, it is easy to interpret students' responses because each choice represents a unique developmental level of student understanding. Second, using OMC questions are more efficient than using traditional multiple-choice questions because they require fewer items to obtain the same amount of data on one's understanding of a construct (Briggs et al. 2010; Briggs and Alonzo 2012).

Furthermore, for this style of multiple-choice test, instead of students choosing only one choice that best reflects their thinking, the assessment items are designed so that students select all the choices that they think are correct, enabling a more precise analysis of students' LP. In order to identify students who select all the choices mechanically without consideration of the assessment items, a common misconception was included among the response choices (Lee 2017). Figure 3.3 shows an example of an OMC item for an assessment on map understanding. The response items correspond to the statements listed in the hypothesized LP. For example, a student with Level 4 understanding (the highest) would select all correct item

Table 3.1 A list of literature reviewed in this study to describe a hypothesized LP for map understanding

Author	Title	Journal name/publication type
Bartz, B. (1970)	Maps in the Classroom	Journal of Geography
Battersby, S. and D. Janelle (2009)	Area Estimation of World Regions and the Projection of the Global-Scale Cognitive Map	Annals of the Association of American Geographers
Blades, M. and C. Spencer (1986)	Map Use by Young Children	Geography
Blades, M. and C. Spencer (1990)	The Development of 3–6 Year-Olds' Map Using Ability: The Relative Importance of Landmarks and Map Alignment	Journal of Genetic Psychology
Boardman, D. (1983)	Graphicacy and Geography Teaching	Book
Boardman, D. (1989)	The Development of Graphicacy: Children's Understanding of Maps	Geography
Gillen, J., L. Skryzhevska, M. Henry and J. Green (2010)	Map Interpretation Instruction in Introductory Textbooks: A Preliminary Investigation	Journal of Geography
Kastens, K. and L. Liben (2007)	Eliciting Self Explanations Improves Children's Performance on a Field-Based Map Skills Task	Cognition and Instruction
Kastens, K. and L. Liben (2010)	Children's Strategies and Difficulties While Using a Map to Record Locations in an Outdoor Environment	International Research in Geographical and Environmental Education
Kim, K. and Y. Kim (2009)	A Study of the Improvement of Graphicacy for Elementary and Secondary Map Education: Based on the Application of Solo Taxonomy Model for Graphicacy Test	Journal of the Korean Cartographic Association
Lee, K. (2007)	Analysis of the Problems of the Maps and the Map Skills in Social Studies Texts of Elementary School	Journal of Geographic and Environmental Education
Liben, L. and R. Downs (1989)	'Understanding Maps as Symbols: The Development of Map Concepts in Children' in 'Advances in Child Development and Behavior'	Book chapter
Liben, L. and R. Downs (1993)	Understanding Person–Space–Map Relations: Cartographic and Developmental Perspectives	Developmental Psychology

(continued)

Table 3.1 (continued)

Author	Title	Journal name/publication type
Liben, L. (2008)	Understanding Maps: Is the Purple Country on the Map Really Purple?	Knowledge Quest
Lobben, A. (2007)	Navigational Map Reading: Predicting Performance and Identifying Relative Influence of Map-Related Abilities	Annals of the Association of American Geographers
Matthews, M. (1984)	Cognitive Mapping Abilities of Young Boys and Girls	Geography
Maxim, G. (1997)	Developmentally Appropriate Map Skills Instruction	Childhood Education
Park, S., H. Kim and E. Lee (2009)	An Analysis of Students' Graphicacy in Korea Based on the National Assessment of Educational Achievement from 2005 to 2007	Journal of the Korean Geographical Society
Schuit, W. (2011)	A Method for Teaching Topographic Map Interpretation	Journal of Geography
Shim, S. (2005)	The Development of Skill-Oriented Learning a Map: Focused on Learning a Scale in Elementary School	The Journal of The Korean Association of Geographic and Environmental Education
Song, U. and J. Lim (2006)	The Process of Learning Picture-Maps for Fostering the Ability of Perceiving Spaces	The Journal of The Korean Association of Geographic and Environmental Education

Source Lee (2017, pp. 144–145)

Table 3.2 The hypothesized LP for map understanding at the stage of specifying construct

Level	Knowledge and skill descriptions	Category
4	• Identify relief features from contour patterns such as valleys, hills, ridges • Understand the effect of location difference between points • Understand characteristics of popular map projections • Calculate approximate areas on maps using distance measures in a map	Contour lines Location Projection Scale
3	• Calculate the average gradient between two points on a map • Understand differences of distance by relief (ups and downs) on a surface • Give locations on atlas maps using latitude and longitude • Plot the eight points of the compass • Understand that all projections cause distortions in varying degrees • Calculate winding distance between two fixed points on a map using a scale bar • Calculate approximate straight-line distance between two fixed points in a map using a scale bar	Contour lines Distance Location Orientation Projection Scale Scale
2	• Read heights from contours on a map and estimate heights between contours • Measure winding distances on maps • Express locations using landmarks • Plot the cardinal directions: North, East, South, and West • Understand that there are many ways to draw a world map, and various types of maps are used depending on the purpose • Understand the features of large- and small-scale maps	Contour lines Distance Location Orientation Projection Scale
1	• Infer elevation based on colors on a map • Measure relative distance on a map (A is closer than B from C) • Describe the relative locations of objects using before, behind, in front of, to the left of, etc • Follow directions using left, right, forwards, in a circle, etc • Consider the bottom of a map as south and top of a map as north • Consider only maps drawn using Mercator projection as accurate and acceptable • Measure straight line distance between two fixed points on a map	Contour lines Distance Location Orientation Orientation Projection Distance

Source Lee (2017, p. 50)

responses in Fig. 3.3, whereas a student with a lower level understanding would select fewer correct item responses.

3.5 Data Collection and Coding

In line with the third step for LP research, a pilot test and a main test were conducted. The pilot test was conducted with 39 middle school and 41 high school students in Korea. In statistical analysis of the LP assessment, the Item Fit Statistics is an indicator of the assessment's validity (Lee 2017; Linacre 2002). Item Fit Statistics

Q. Table 1 shows countries around the world. Consider whether the countries listed in this table are accurately reflected in the following two world maps. Next, select all of the following explanations about these maps that you think are right.

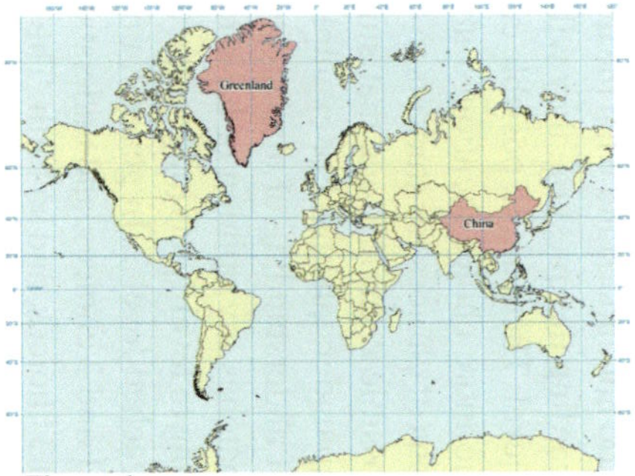

Figure 4

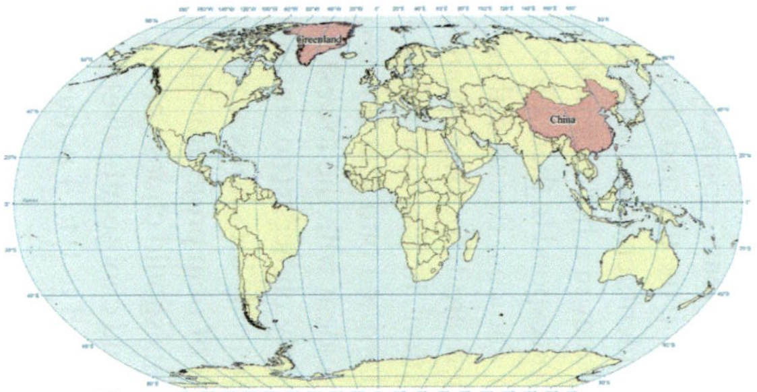

Figure 5

Table 1.

Rank	Country	Area (km²)
1	Russia	17,075,200
2	Canada	9,984,670
3	USA	9,826,630
4	China	9,596,960
5	Brazil	8,511,965
⋮	⋮	⋮
12	Greenland	2,166,086
13	Saudi Arabia	2,149,690
⋮	⋮	⋮

① The world map in Figure 5 is correct, and both maps can be used, depending on the intended purpose.
② Figures 4 and 5 are maps drawn by distorting one or more elements of each country's shape, distance, area, and orientation.
③ It is better to use the map in Figure 5 rather than Figure 4 to compare the relative sizes of the countries in the world.
④ The map in Figure 4 is the only accurate world map without any distortions on shape, area, or distance.
⑤ No statement matches my opinion or have no idea.

Fig. 3.3 An example of OMCs used in this case study (*modified in* Lee 2017, pp. 150–151)

Table 3.3 Demographic characteristics of participants

Participants	Gender		Total
	Male	Female	
3rd graders (9 years old)	176 (49.0%)	183 (51.0%)	359 (100.0%)
5th graders (11 years old)	80 (47.6%)	88 (52.4%)	168 (100.0%)
7th graders (13 years old)	172 (49.3%)	177 (50.7%)	349 (100.0%)
10th graders (16 years old)	168 (48.4%)	179 (51.6%)	347 (100.0%)
College students (19–23 years old)	123 (46.8%)	140 (53.2%)	263 (100.0%)
Total	719 (48.4%)	767 (51.6%)	1486 (100.0%)

Source Lee (2017, p. 72)

uses mean square residual (MNSQ) and standardized mean square residual (ZSTD) to show whether each assessment item accurately measures what it intended to measure (Bond and Fox 2007). If MNSQ and ZSTD are within a certain range, it can be considered that the validity of the test has been verified. As the MNSQ and ZSTD of the pilot test are located within the proper range, the pilot test was confirmed as valid.

The main test was administered to 1,486 students, comprising 359 3rd graders, 168 5th graders, 349 7th graders, 347 10th graders, and 263 tertiary students. The demographic information of the participants is presented in Table 3.3.

To measure students' competency in understanding maps, students were asked to read the choices in the test and choose the ones they thought were correct. This test is designed so students choose choice ⑤if they do not understand what the test item asks. If students have even a little knowledge regarding the question, they will choose the choices that correspond to their knowledge level. There are 15 possible combinations from the four choices. Because there is one more choice mentioned above (⑤I do not know), there are 16 possible combinations of choices for students to choose from for each question. In this study, the 16 possible combinations that students could select were classified into four levels of map understanding. For example, as seen in Fig. 3.3, the student with the lowest level of understanding of map projection would have chosen only choice ⑤(no statement matches my opinion or have no idea). If a student chooses only the choice corresponding to level 1, it is judged that the student's understanding of map projection has not reached level 2. In general, understanding Level 2 students should choose both Level 1 and 2 choices, and understanding Level 3 students should choose all Level 1, 2, and 3 choices, and so forth. The result of the test was converted to 0 to 4 points according to Table 3.4, which shows the scoring system in general.

Table 3.4 Scoring rubric for map understanding test (map projection)

Score	Choices selected by students
4	Level 1, Level 2, Level 3, Level 4
3	Level 1, Level 2, Level 3
2	Level 1, Level 2
1	Level 1
0	Nothing or Choice ⑤

However, for the example in Fig. 3.3, there is one more thing to consider. The Level 1 choice presented in Fig. 3.3 represents a misconception related to map projection. When a student has a misconception, it means that the student does not fully understand the concept, but partially understands it. Therefore, if a student chooses a misconception, the student should not be regarded as having a complete understanding of the concept.

Moreover, if a student does not select answers that correspond to prior developmental stages and instead chooses one or two higher level choices to be correct (i.e., selects only one Level 3 choice or selects Level 2 and Level 4 choices) it is not easy to judge the understanding level of the student. To solve this dilemma, five reviewers in geography education (three graduate students, one middle school geography teacher, and one high school geography teacher with more than five years' experience) were invited to judge the level of students' understanding through discussion based on the students' responses.

3.6 Data Analysis—The Measurement Model

The partial credit Rasch model (Masters 1982; Rasch 1980) was used to analyze the test data in this chapter. The Rasch model is the most widely used analysis tool in LP research because it enables researchers to numerically analyze the correlation between test item difficulty and student competency (Lee 2016). Thus, it is more reliable than traditional measurement theories (Liu and Boone 2006). A partial credit Rasch model is needed when analyzing data that scored students' responses or behaviors using scoring rubrics (Liu 2010). It is appropriate to apply this model in this study because the test items were not designed in a way that bases scores on the correctness of the students' responses. Each choice in the test items signifies the developmental stages of map understanding, and the test was designed to have the students select the choices they thought were right. The students' responses from the test were scored from 0–4 using a rubric. ConstructMap software v4.6.0 (Kennedy et al. 2008) was used to derive students' LP by applying the partial credit Rasch model.

The Item Characteristic Curve (ICC) shows the difficulty of the test items and the percentage of correct answers as logit values. Research in education often uses the ICC to analyze item validity, student characteristics, and item difficulty (Ji and Chae 2000). In LP research, the ICC can be used to examine the item validity as it shows the correlation between the student ability and the correct answer rate. The shape of the curve gives information about whether the item is measuring what it proposes to measure. The ConstructMap creates the ICC using the test data. In statistics, the logit is a function that creates a map of probability values from 0 to infinity (Hilbe 2009). Since the statistics are beyond the purpose of this chapter, in this study, you may think that a low logit value is an easy concept to understand, or the students' level is low and vice versa.

In the legend, p(0) represents the lowest level and p(4) represents the highest level of skills or understanding in map projection. The curve of p(0) shows that the probability of choosing Level 0 in the question item is high when a respondent's ability is low, and the probability decreases as the respondent's ability increases. On the other hand, the curve of p(4) shows that the probability of choosing Level 4 increases as the respondent's ability increases. For example, in Fig. 3.4, a student at the proficiency of -1.0 is most likely to show Level 1 understanding because the probability of choosing Level 0 (p(0)) is about 0.33, Level 1 (p(1)) is about 0.48, Level 2 (p(2)) is about 0.17, Level 3 (p(3)), and Level 4 (p(4)) are both about 0. In the present study this means the student is most likely to choose Level 1 choice 4—'The map in Fig. 4 is the only accurate world map without any distortions on shape, area, or distance' because it has the highest probability of 0.48. Recall that the hypothesized LP for map projection Level 1 is 'Consider only maps drawn using Mercator projection as accurate and acceptable'.

In the Rasch model, the difficulty level at which a student has a 50% chance of choosing the correct answer is defined as the student's proficiency level. For example, if a student's proficiency is 1.0, the probability that this student will choose the level 2 choice is 0.5. Then, the proficiency at this point will be the difficulty (1.0) of the level 2 choice. Thus, students with a higher proficiency than the difficulty of the level 2 choice will feel that the answer is easy, and students with a lower proficiency than the choice will feel that the answer is difficult.

In Fig. 3.4, there are points where each curve intersects. Among them, the parts where adjacent developmental levels meet, such as p(0) and p(1), p(1) and p(2), p(2) and p(3), and p(3) and p(4), are the points at which the developmental level changes (Ji and Chae 2000; Wu and Adams 2007). The ICCs of all six items in the map understanding test were created in the same way. Based on the ICC, students' LP for map understanding was derived by calculating the probability of students' selection of a choice for each question.

The ConstructMap programme generates a Wright Map, which is a diagram based on calculated proficiency of students and the difficulty of each OMC item. Figure 3.5 shows a Wright Map for all grade groups—middle, high school, and college students.

The Wright Map shows a diagram of the frequency distribution by placing the ability of the students on the same assessment scale as the difficulty of the test items (Kennedy et al. 2008; Lee 2017). The left side shows the distribution of students'

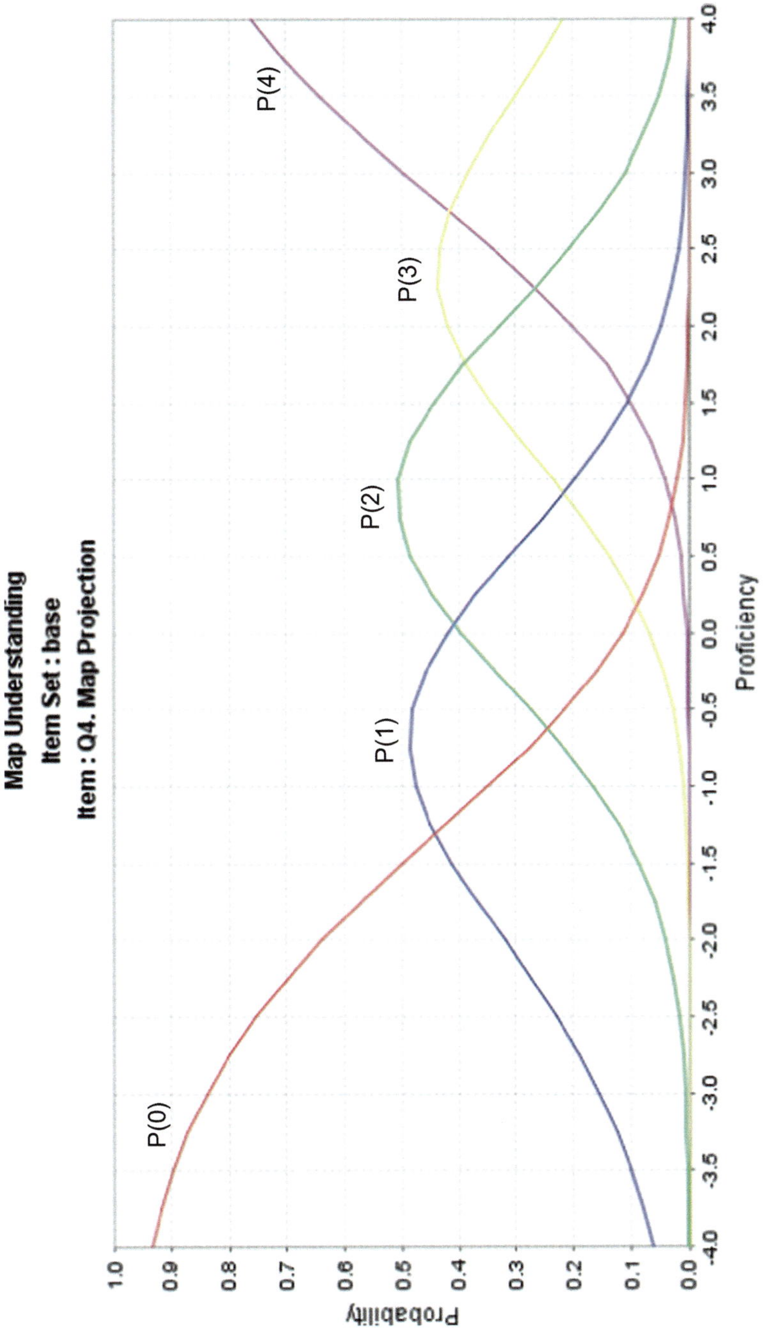

Fig. 3.4 Item characteristic curve for the example question item: Map projection (*Source* Lee 2017, p. 88)

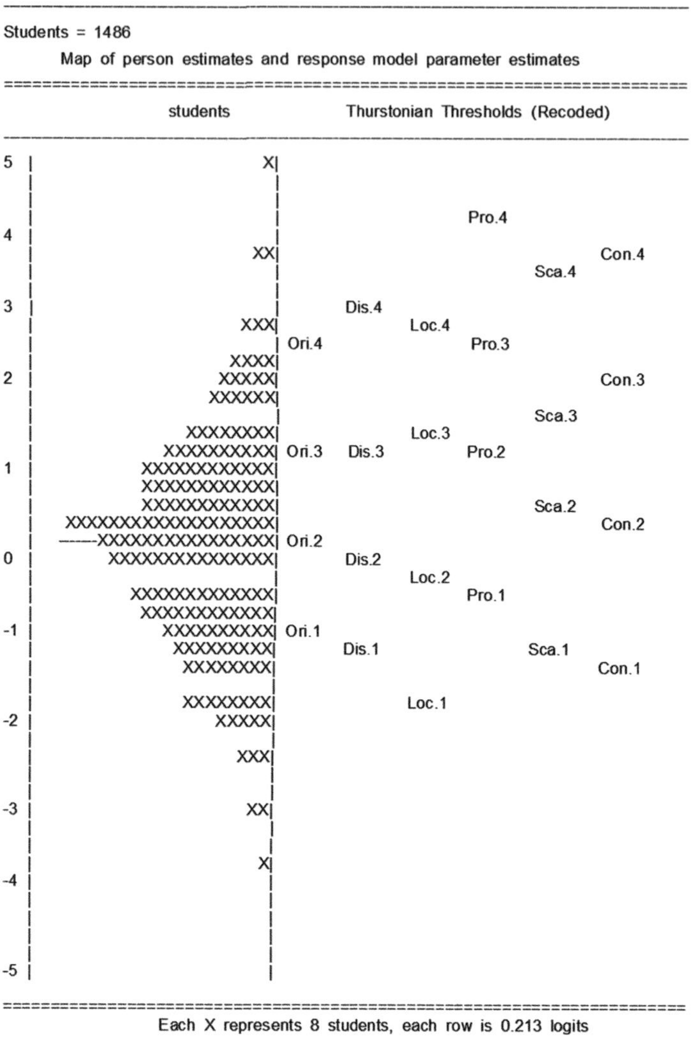

Fig. 3.5 All students' Wright Map for map understanding (*modified in* Lee 2017, p. 126)

proficiency, and the right side shows the distribution of test item choices that reflect the difficulty of developmental stage. The y axis represents students' proficiency and test item difficulty as logit values. Each 'X' on the left side represents the distribution and number of students. From the bottom of the map to the top, a student's ability level increases, and the '——' line near the −1 logit score indicates a student's average proficiency level. The Thurstonian threshold on the right side of the map represents a student's proficiency (logit value) when the probability of choosing the correct answer on a test item is 0.5. That is, the difficulty of the developmental levels suggested by the hypothesized LP is distributed on the right side of the map as the Thurstonian threshold. In Fig. 3.5, the Level 1 choice in 'Location' (shown as 'Loc.1') item is the easiest to grasp, and the Level 4 choice in 'Map Projection' (shown as 'Pro.4') item is the most difficult to understand. Using the Wright map makes it possible to understand both the degree of difficulty and the development sequence of knowledge and skills related to map understanding, and students' degree of proficiency. The Wright Map is also used to identify actual LP of students (Maeng et al. 2013). In this way, the Wright Map is used as the basis for correcting the hypothesized LP (Table 3.2) by reflecting the results of the students' OMC assessment item responses. In addition, by analyzing the Wright Map, the choices of the OMC assessment items can be improved and made suitable for the students' development level.

A final LP for map understanding was developed by analyzing the Wright Map as shown in Table 3.5. The sequence is the same as the common developmental process assumed in the hypothesized LP. However, differences between the hypothesized LP and the actual LP were found in the degree of difficulty of knowledge and skills related to map understanding. For example, it was hypothesized that the difficulty of what was identified as ability Level 3 in contour lines, 'Calculate the average gradient between two points on a map', was actually found to be Level 4 (Lee 2017). The knowledge and skill descriptions in regular text in Table 3.5 are those with no changes between the hypothesized LP and the identified LP (in terms of difficulty). The descriptions marked in bold text proved to be more difficult than assumed, and those in italicized text proved to be easier than assumed.

Analysis of the results revealed that there is a sequence of learning that students commonly experience, and the abilities related to map elements can be categorized into simple concepts and complex concepts according to difficulty levels. In the present research located in Korea, content related to orientation, distance, and location are relatively easy for students to learn, whereas scale, contours, and map projections are relatively difficult.

Identifying whether a curriculum is aligned with students' developmental stage (LP) is essential to assess whether the learning objectives are appropriate and whether the content and processes proposed to achieve the learning objectives are suitable. For example, it was found that the geography curriculum in Korea is not well-sequenced to students' map understanding, usually taught in 3rd–4th grade. For LP research to fulfill its role successfully, the development of effective assessment tools is important (Lee 2017). The assessment triangle covered in Sect. 3.2.1 is widely used as a methodological framework in LP research. Of the three elements of the assessment triangle consisting of observation, cognition, and interpretation, the fact that two elements of observation and cognition are related to the development of an effective assessment (Maeng et al. 2013) shows the importance of assessment in LP research. In addition, statistical methods such as Rasch model or Wright Map analysis, which are required to derive LPs, cannot play their role if the assessment that measures students' LP does not work properly. Therefore, in order to accurately derive the students' LP, longitudinal follow-up studies over a relatively long period of time and cross-sectional studies comparing achievements between grades should be conducted in parallel using various assessments (Maeng et al. 2013). Since it is very difficult

Table 3.5 The identified learning progression on map understanding among Korean students

	ContourlLines	Distance	Location	Orientation	Projection	Scale
Difficulty level 4	• Identify relief features from contour patterns such as valleys, hills, ridges • **Calculate the average gradient between two points on a map**	• **Understand differences of distance by relief (ups and downs) on a surface**	• Understand the effect of location difference between points	• **Plot the eight points of the compass**	• Understand characteristics of popular map projections • **Understand the fact that all projections cause distortions in varying degrees**	• Calculate approximate areas on maps using distance measures on a map
Difficulty level 3		• **Measure winding distances on maps**	• Give locations on atlas maps using latitude and longitude	• Plot the cardinal directions NESW	• **Understand the fact that there are many ways to draw a world map, and various types of maps are used depending on the purpose of use**	• Calculate winding distance between two fixed points on a map using a scale bar
Difficulty level 2	• Read heights from contours on a map and estimate heights between contours	• **Measure straight line distance between two fixed points on a map**	• Express their locations using landmarks near them	• **Regard bottom of a map as south and top of a map as north**	• **Regard only maps drawn using Mercator projection as accurate and acceptable**	• Calculate approximate straight line distance between two fixed points on a map using a scale bar

(continued)

Table 3.5 (continued)

	ContourlLines	Distance	Location	Orientation	Projection	Scale
Difficulty level 1	• Infer elevation based on colors on a map	• Measure relative distance on a map (A is closer than B from C)	• Describe the relative locations of objects using before, behind, in front of, to the left of, etc	• Follow directions using left, right, forwards, in a circle, etc		• *Understand the features of large- and small-scale maps*

Note Regular text = No change between hypothesized LP and the identified LP in terms of difficulty; **Bold text** = more difficult than assumed; *Italicized text* = easier than assumed
Source Lee (2017, p. 109)

for one researcher to develop various assessment tools and derive LPs, many geography education researchers need to pay attention to develop more diverse and better assessments.

3.7 Conclusion

Recent interest in LP research has been driven by effective science education curriculum that is well-aligned with students' developmental processes. Geography educators who are concerned with LP research contend that developing appropriate methodologies to better understand student LPs on geographic concepts and skills is imperative. This chapter introduced a process of identifying LPs with an example of map understanding. First, a hypothesized LP on map understanding based on institutional and academic literature was established. Next, the map understanding test, an assessment tool to measure actual LP for students' map understanding was developed, based on the hypothesized LP. The main test results then generated quantitative information, such as ICCs and a Wright Map.

The findings from this study suggest that with an appropriate assessment, a curriculum aligned with students' developmental stages and appropriate learning strategies can be developed and established. In LP research, you need to develop an assessment that can accurately measure students' LP. From multiple-choice assessments to observations and interviews, assessments that can accurately measure the student's developmental stage should be developed and utilized. Through this, it is possible to know students' development level on geographical concepts more accurately, to evaluate whether the current curriculum is suitable for students' developmental level, and to obtain data for designing plans for better geography education. LP research is not completed by applying a one-time evaluation process, though. Continuous and repetitive evaluation of the assessment, interpretation of the assessment result, and validation of the derived LP are critical. Through this process, it becomes possible to precisely grasp students' LP on a geographic concept or skill of interest.

References

Alonzo A, Steedle J (2009) Developing and assessing a force and motion learning progression. Sci Educ 93(3):389–421

Bartz B (1970) Maps in the classroom. J Geogr 69(1):18–24

Battersby S, Janelle D (2009) Area estimation of world regions and the projection of the global-scale cognitive map. Ann Assoc Am Geogr 99(2):273–291

Bednarz S, Heffron S, Huynh N (2013) A road map for 21st century geography education. Geography Education Research, Washington, DC

Blades M, Spencer C (1986) Map use by young children. Geogr 71(1):47–52

Blades M, Spencer C (1990) The development of 3 to 6 year-olds' map using ability: the relative importance of landmarks and map alignment. J Genet Psychol 151(2):181–194

Boardman D (1983) Graphicacy and geography teaching. Croom Helm, London

Boardman D (1989) The development of graphicacy: children's understanding of maps. Geogr 74(4):321–331

Bond T, Fox C (2007) Applying the Rasch model: fundamental measurement in the human sciences. Lawrence Erlbaum Associates, Mahwah

Briggs D, Alonzo A (2012) The psychometric modeling of ordered multiple-choice item responses for diagnostic assessment with a learning progression. In: Alonzo AC, Gotwals AW (eds) Learning progressions in science: current challenges and future directions. Sense Publishers, Rotterdam, pp 293–316

Briggs D, Alonzo A, Schwab C, Wilson M (2010) Diagnostic assessment with ordered multiple-choice items. Educ Assess 11(1):33–63

Chang CH, Tan GCI, Tan J, Kwek CH (2020) Developing a learning progression for climate change in geography education. Technical Report. Office of Education Research. National Institute of Education, Singapore

Clements D, Wilson D, Sarama J (2004) Young children's composition of geometric figures: a learning trajectory. Math Think Learn 6(2):163–184

Corcoran T, Mosher F, Rogat A (2009) Learning progressions in science: an evidence-based approach to reform. Consortium for Policy Research in Education, Philadelphia

Duncan R, Hmelo-Silver CE (2009) Learning progressions: aligning curriculum, instruction, and assessment. J Res Sci Teach 46(6):606–609

Duschl R, Maeng S, Sezen A (2011) Learning progressions and teaching sequences: a review and analysis. Stud Sci Educ 47(2):123–182

Gillen J, Skryzhevska L, Henry M, Green J (2010) Map interpretation instruction in introductory textbooks: a preliminary investigation. J Geogr 109(5):181–189

Hilbe J (2009) Logistic regression models. Chapman & Hall/CRC, Boca Raton

Huynh N, Solem M, Bednarz S (2015) A road map for learning progressions research in geography. J Geogr 114(2):69–79

Ji E, Chae S (2000) Theories and practices of the Rasch modeling. Educational Science, Seoul

Johnson P, Tymms P (2011) The emergence of a learning progression in middle school chemistry. J Res Sci Teach 48(8):849–877

Kastens K, Liben L (2007) Eliciting self explanations improves children's performance on a field-based map skills task. Cognit Instruct 25(1):45–74

Kastens K, Liben L (2010) Children's strategies and difficulties while using a map to record locations in an outdoor environment. Int Res Geogr Environ Educ 19(4):315–340

Kennedy C, Wilson M, Draney K, Tutnciyan S, Vorp R (2008) ConstructMap v4.4.0: quick start guide. BEAR Center, Berkley
Kim K, Kim Y (2009) A study of the improvement of graphicacy for elementary and secondary map education: based on the application of solo taxonomy model for graphicacy test. J Kor Cartographic Assoc 9(1):45–61
Lee H (2016) An application of item response theory to investigate the validity of a learning progression for number sense. University of California, Berkeley
Lee H, Liu O (2010) Assessing learning progression of energy concepts across middle school grades: the knowledge integration perspective. Sci Educ 94:665–688
Lee J (2017) Learning progressions on map understanding: a case study of elementary through college students in Korea. Texas State University, San Marcos
Lee K (2007) Analysis of the problems of the maps and the map skills in social studies texts of elementary school. J Geogr Environ Educ 15(2):109–123
Liben L (2008) Understanding maps: is the purple country on the map really purple? Knowl Quest 36(4):20–30
Liben L, Downs R (1989) Understanding maps as symbols: the development of map concepts in children. In: Reese HW (ed) Advances in child development and behavior. Academic Press, New York, pp 145–201
Liben L, Downs R (1993) Understanding person–space–map relations: cartographic and developmental perspectives. Dev Psychol 29:739–752
Linacre J (2002) What do infit and outfit, mean-square and standardized mean? Rasch Meas Trans 16(2):878
Liu X (2010) Using and developing measurement instruments in science education: a Rasch modeling approach. Information Age Publishing, Charlotte
Liu X, Boone W (2006) Introduction to Rasch measurement in science education. In: Liu X, Boone W (eds) Applications of Rasch measurement in science education. JAM Press, Maple Grove, pp 1–22
Lobben A (2007) Navigational map reading: predicting performance and identifying relative influence of map-related abilities. Ann Assoc Am Geogr 97(1):64–85
Maeng S, Lee K (2015) Cross-sectional item response analysis of geocognition assessment for the development of plate tectonics learning progressions: Rasch model. J Kor Assoc Sci Educ 35(1):37–52
Maeng S, Seong Y, Jang S (2013) Present states, methodological features, and an exemplar study of the research on learning progressions. J Kor Assoc Sci Educ 33(1):161–180
Masters G (1982) A Rasch model for partial credit scoring. Psychometrika 47(3):149–174
Matthews M (1984) Cognitive mapping abilities of young boys and girls. Geogr 69(4):327–336
Maxim G (1997) Developmentally appropriate map skills instruction. Child Educ 73(4):206–211
Mislevy R, Steinberg L, Almond R (2003) On the structure of educational assessments. Measurement 1(1):3–62
National Research Council (2001) Knowing what students know: the science and design of educational assessment. The National Academies Press, Washington, DC
National Research Council (2006) Systems for state science assessment. National Academy Press, Washington, DC
National Research Council (2007) Taking science to school: learning and teaching science in grades K–8. The National Academies Press, Washington, DC
National Research Council (2012) A framework for K–12 science education: practices, crosscutting concepts, and core ideas. The National Academies Press, Washington, DC
Park S, Kim H, Lee E (2009) An analysis of students' graphicacy in Korea based on the national assessment of educational achievement from 2005 to 2007. J Kor Geogr Soc 44(3):410–427
Pellegrino J (2009) The challenges of conceptualizing what low achievers know and how to assess their competence. In: Perie M (ed) Considerations for the alternate assessment based on modified achievement standards (AA-MAS): understanding the eligible population and applying that

knowledge to their instruction and assessment. New York Comprehensive Center, New York, pp 90–151
Plummer J, Krajcik J (2010) Building a learning progression for celestial motion: elementary levels from an Earth-based perspective. J Res Sci Teach 47(7):768–787
Rasch G (1980) Probabilistic models for some intelligence and attainment tests. University of Chicago Press, Chicago
Schuit W (2011) A method for teaching topographic map interpretation. J Geogr 110(5):209–216
Shim S (2005) The development of skill-oriented learning a map: focused on learning a scale in elementary school. J Kor Assoc Geogr Environ Educ 13(2):263–274
Solem M, Huynh N, Boehm R (2014) Learning progressions for maps, geospatial technology and spatial thinking: a research handbook. Association of American Geographers, Washington, DC
Song U, Lim J (2006) The process of learning picture-maps for fostering the ability of perceiving spaces. J Kor Assoc Geogr Environ Educ 14(2):119–134
Songer N, Kelcey B, Gotwals A (2009) How and when does complex reasoning occur? Empirically driven development of a learning progression focused on complex reasoning about biodiversity. J Res Sci Teach 46(6):610–631
Wu M, Adams R (2007) Applying the Rasch model to psycho-social measurement: a practical approach. Educational Measurement Solutions, Melbourne

Jinhee Lee is an Assistant Professor of Geography Education at Pusan National University in Korea. He teaches courses in geography education, including theories in geography education, curriculum development, and evaluation for geography pre-service and in-service teachers. His research focuses on geography education, learning progression, spatial thinking education, and education policy.

Injeong Jo an Associate Professor of Geography and Environmental Studies at Texas State University. Her research focuses on geography and spatial thinking education, geospatial technologies for teaching and learning, as well as teacher education. She teaches courses in geographic education for both undergraduate and graduate students, including Fundamental Themes in Geography, Spatial Thinking in Education, and Technology in Geographic Education. She currently serves as the Faculty Coordinator of the Online Master of Applied Geography in Geographic Education Program.

Chapter 4
Assessing Spatial Skills/Thinking in Geography

Jongwon Lee and Injeong Jo

Abstract The acquisition of spatial skills/thinking is essential for both everyday functioning and academic purposes; therefore, valid and reliable assessment of this capability is an important task. However, to date, there have been few assessment instruments developed for spatial skills/thinking, and measures and approaches vary across disciplines and from study to study. The goal of this chapter is to review and synthesize peer-reviewed articles on the measurement/assessment of spatial skills/thinking. A search informed by systematic literature review protocols was conducted using Web of Science and Google Scholar as well as a manual search of well-known journals. A total of 19 papers were selected for full review with four broad trends in research identified: (1) measuring spatial skills/thinking on a small scale without any specific context, (2) measuring spatial skills/thinking on an environmental scale in everyday life contexts, (3) measuring spatial skills/thinking on a geographic scale in geoscience contexts, and (4) spatial skills/thinking self-assessments in everyday life contexts. This review broadens the understanding of spatial skills/thinking assessment across disciplines and provides directions for future research.

Keywords Spatial skills · Spatial thinking · Scale · Assessment instruments

4.1 Introduction

Spatial thinking is a skill applicable to many subject areas, ranging from psychology and science to geography. Each discipline emphasizes different aspects of spatial skills/thinking. For example, while psychologists see spatial ability as a factor of human intelligence and focus on small-scale spatial skills such as mental rotation,

J. Lee (✉)
Ewha Womans University, Seoul, South Korea
e-mail: jongwonlee@ewha.ac.kr

I. Jo
Texas State University, San Marcos, TX, USA

T. Bourke et al. (eds.), *Assessment in Geographical Education: An International Perspective*, Key Challenges in Geography,
https://doi.org/10.1007/978-3-030-95139-9_4

geographers consider spatial thinking at larger scales and focus on patterns and interconnections (Lee and Bednarz 2012). Accordingly, methods for assessing or measuring these skills across disciplines are so different that a comparison of the findings is often problematic.

The National Research Council's (NRC) (2006) report *Learning to Think Spatially* presented a new interdisciplinary approach to spatial skills/thinking. It defined spatial thinking in three categories—knowledge; tools and skills; and habits of mind—which are a collection of cognitive skills comprised of knowing concepts of space, using tools of representation, and reasoning processes. Taken together these three categories give spatial skills/thinking its power of versatility and applicability. However, since the report focused on the conceptualization of spatial skills/thinking and support systems to develop spatial thinking, the lack of reliable and valid assessment of spatial thinking still remains a problem. The report argued 'there are neither content standards nor valid and reliable assessments for spatial thinking. Without such standards and assessments, spatial thinking will remain locked in a curious educational twilight zone' (NRC 2006, p. 232). This assertion from the NRC emphasized that more research should be conducted on the assessment or measurement of spatial skills/thinking. This chapter responds to this assertion by presenting a review of recent developments in spatial skills/thinking measurement/assessment. The chapter has four sections. First, we outline tensions in definitions and conceptualizations from the various disciplines that use spatial skills/thinking and how it is measured or assessed. Then we outline the methods for the current study. Next, the findings are presented as four main groups based on their conceptual features, which we critically review and synthesize according to current status, trends, and challenges in spatial thinking measurement/assessment across disciplines. Lastly, we conclude the review with future directions in the field.

4.2 Definitional and Conceptual Tensions

Spatial thinking is often used synonymously with spatial ability, especially in the field of psychology. A number of prevailing definitions exist for spatial ability. For example, one of the most widely adopted definitions of spatial ability describes it as the 'ability to generate, retain, retrieve, and transform well-structured visual images' (Lohman 1996, p. 112). Sutton and Allen (2011) defined it as 'the performance on tasks that require: (a) the mental rotation of objects; (b) the ability to understand how objects appear in different positions; and (c) the ability to conceptualize how objects relate to each other in space' (p. 5). The concept often involves elements such as holding a visual pattern in memory, comparing visual patterns, or doing a mental transformation, requiring the manipulation of internal (mental) representations. They derive in part from the psychometric tradition of intelligence measurement and testing with, for example, a test kit that identifies types of spatial ability, such as spatial orientation and visualization (Ekstrom et al. 1976).

Spatial abilities are not a single unified construct, and it is less than clear what the most effective ways are to categorize and measure different types of spatial skills (Voyer et al. 1995). Researchers use different explanations for the mental processes used in solving spatial problems, and this affects their definitions of the components of spatial ability. The commonly listed components include spatial visualization, spatial orientation, and mental rotation (Carroll 1993; Linn and Petersen 1985; McGee 1979) which are now elaborated.

Spatial visualization is an ability to mentally manipulate, rotate, twist, and pictorially invert presented visual stimuli (McGee 1979). The underlying ability involves a process of recognition, retention, and recall of a configuration in which there is movement among the internal parts of the configuration, or of an object manipulated in three-dimensional space, or the folding or unfolding of flat patterns. Spatial visualization tasks include paper folding, surface development, block design, and Guilford-Zimmerman's spatial visualization (Linn and Petersen 1985).

Spatial orientation is the ability to imagine objects from different perspectives (Turgut 2015). It is the notion of perspective taking, and in a spatial orientation task, one has to mentally or physically position himself or herself in the place of the object to be manipulated to determine the position of the object or the result of a transformation on the object.

Mental rotation is a cognitive process in which a person imagines how two-dimensional and three-dimensional objects would appear after they have been rotated around a point at a certain angle (Shepard and Metzler 1988). The Vandenberg and Kuse test (Vandenberg and Kuse 1978) is a commonly used instrument to measure mental rotation. Subjects are asked to compare three-dimensional figures to decide whether they are similar.

Contrary to the broad characterization of spatial ability, however, research has largely focused on measuring only small-scale and object-based spatial ability (Bednarz and Lee 2019). Traditional attempts to measure spatial ability have not paid attention to scale (Nazareth et al. 2019) probably because psychometric tests do not measure large-scale skills. Considering that Hegarty and Waller (2004), who reviewed previous studies on spatial abilities, found very little correlation between large- and small-scale spatial abilities, it is important to recognize the impact of factor of scale differences (Bednarz and Kemp 2011). That is to say that someone who is good at spatial tasks on a small scale is not necessarily good at large-scale spatial tasks. Montello (1993) discusses the importance of scale in understanding psychological spaces and defines four classes of psychological spaces based on the *projective* size of the space relative to the human body rather than its absolute size—figural, vista, environmental, and geographical. Small-scale spatial skills are needed on the figural (e.g., a small object) and vista (e.g., a single room) scales, where an individual can visually observe all spatial characteristics from a single vantage point. In contrast, large-scale spatial skills are necessary on the environment scale (e.g., neighborhoods), where an individual must move to obtain information about spatial properties of the space, and on a geographical scale (e.g., country), where spatial information can be acquired only through symbolic representation such as map.

Geographers have argued that a psychological conceptualization of spatial ability overlooks some important aspects of spatial phenomena such as distribution, process, association, and structure (Golledge 1993; Self and Golledge 1994), and only sees spatial skills/thinking as operating at a small scale (Liben 1981). Golledge and Stimson (1997), among others, argued for the addition of an ability termed 'spatial relation' composed of a collection of spatial skills including abilities to recognize spatial distributions and spatial patterns, to connect locations, to associate and correlate spatially distributed phenomena, to comprehend and use spatial hierarchies, to regionalize, to orientate to real-world frames of reference, to imagine maps from verbal descriptions, to sketch maps, to compare maps, and to overlay and dissolve maps. Gersmehl and Gersmehl (2006) also offered a taxonomy of spatial thinking defined as skills that geographers use to analyze the spatial relationships in the world. The range of skills identified are mostly used on a geographic scale (e.g., map interpretation, geographic analysis, etc.). At the same time, geography educators have tried to develop assessments that incorporate spatial concepts suggested by these geographers (e.g., Lee 2005; Lee and Bednarz 2009).

The NRC report on *Learning to Think Spatially* (NRC 2006) is the result of a multi-year collaboration among experts in the fields of education, geography, earth and planetary sciences, psychology, and cognitive science. It provides an overview of the value of spatial thinking across the curriculum and presents a general framework for the nature and functions of spatial thinking. The NRC (2006) used the term 'spatial thinking' rather than spatial ability and mentioned that spatial ability is related to but much more restricted in its scope than spatial thinking. According to them, spatial thinking is one form of thinking, which can be distinguished from other forms of thinking, such as verbal, logical, and mathematical thinking, and is not a single ability or process but rather a combination of concepts of space, tools of representation, and processes of reasoning. For example, to think spatially, one must develop a knowledge of spatial concepts such as direction, distance, and spatial association; attain skills in constructing and interpreting graphic representations such as diagrams, maps, and graphs; and acquire and practice the cognitive strategies (or habits of mind) that facilitate problem-solving and decision-making in spatial contexts. The NRC (2006) also claimed that spatial thinking is useful in three different contexts: everyday life, the workplace, and science. In these contexts, spatial thinking can be used to structure problems, find answers, and express solutions.

In everyday living (Golledge 2003), spatial thinking could involve using a map to guide you through an unfamiliar city, merging into high-speed traffic, or orienting yourself in an environment such as learning your way around a new school building. Second, from a vocational or workplace perspective, spatial thinking is an inherent requirement for various professions, such as architects, engineers, town planners, or surveyors. Advances in spatial technologies and the increasing availability of spatial data and low-cost data collection, analysis, and representation tools have contributed to the application of spatial thinking in all areas of social science (Hespanha et al. 2009). Lastly, spatial thinking reliably predicts scholastic and professional success and career choices, particularly in science, technology, engineering, and mathematics (STEM) and related fields (Alias et al. 2002; Jones and Burnett 2008; Marunic and Glazar 2013; Small and Morton 1983; Wai et al. 2009). In fact, several correlational studies suggest a relationship between STEM achievement and spatial thinking—high STEM achievers tend to have high spatial thinking levels (Wang and Carr 2020; Xie et al. 2020).

However, there exists a notable gap between the assessment studies of spatial ability and spatial thinking. In contrast to the long tradition of psychometric testing of spatial ability, formal methods for assessing spatial thinking have not yet been sufficiently developed (Eliot and Czarnolewski 2007; Huynh and Sharpe 2013). Bednarz and Kemp (2011) have argued for attention to the ways that spatial literacy can be measured and evaluated so that powerful, reliable, authentic, and valid assessment instruments can be devised for teaching, learning, and research.

Difficulties in assessing spatial thinking are due to two characteristics of spatial thinking: interdisciplinary and multi-scale nature. The research studies of spatial thinking originate in many different fields of study (e.g., geography education, cognitive psychology, learning sciences, and neurosciences) and thus emphasize different elements of spatial thinking. That is, spatial thinking research acknowledges that concepts of space not only draw from multiple disciplines but also that the application of spatial concepts is entwined with discipline-specific practices. In many cases, the specificity of the task and the context in which it was measured prevents findings from being generalized (Bednarz and Kemp 2011; Hegarty and Waller 2004; Wakabayashi and Ishikawa 2011). Another point of difficulty is that different disciplines use different scales of spatial thinking. For example, cognitive psychologists have focused spatial skills on table-top and computer-generated tasks to understand spatial visualization and orientation, while many geography education researchers focus on spatial representation at a geographical scale. Neuroscientists tend to focus more closely on aspects of brain functionality as it relates to performing spatial thinking tasks (Bednarz and Lee 2019). Therefore, to understand and measure spatial thinking in one big picture, multiple scale approaches are essential. The next section outlines the methods used in the current study.

4.3 Methods

As this chapter is not a systematic review, we conducted the search strategies to ensure that we had located and read as many relevant papers as possible before conducting the critical review. The following steps were adhered to: (1) search strategy including inclusion/exclusion criteria; (2) data extraction and analysis; and (3) presentation of the findings.

4.3.1 Search Strategy

A search informed by systematic literature review protocols was conducted of peer-reviewed papers relating to the development of spatial skills/thinking measures published since 2006. We chose this timeframe to provide a contemporary overview of spatial skills/thinking assessment since the seminal publication *Learning to Think Spatially* (NRC 2006). We based the search on works contained in Web of Science and Google Scholar databases. The search protocol was complemented with a manual search of relevant journals based on keywords, including 'spatial thinking', 'spatial ability', and 'spatial skills'. By way of data evaluation, we first read the abstracts of these papers to see if they were relevant. Second, the full text of the relevant papers were then read. Inclusion criteria focused on the development of spatial skills/thinking assessment. We excluded studies on assessment more broadly or studies that modified measures rather than presenting original ideas. We also examined the reference lists of relevant papers to ensure if any further papers could be included. Finally, a total of 19 papers were identified for review.

4.3.2 Data Extraction and Analysis

When all the papers from database searches were read in full, the following information was extracted into a table for analysis (Table 4.1): (a) background—fields of research, intended use of the measure; (b) scales and types of spatial skills/thinking; small scale (e.g., spatial visualization, spatial orientation, mental rotation), environmental scale (e.g., wayfinding), and geographic scale (e.g., spatial relations); (c) context of spatial skills/thinking—spatial thinking in everyday, spatial thinking at work, spatial thinking in geoscience, and context free; and (d) measurement mode—response format. We repeatedly read the papers to identify key themes and concepts and to identify areas of similarity and difference between the included studies.

4.4 Findings and Discussion

Table 4.1 reveals the studies that were analyzed for this review including information on scale, context of spatial thinking, and response format for each study. As shown in the table, four common research trends emerged. These trends include:

1. Group A: Measuring spatial skills/thinking on a small scale without any specific context [articles 1–8]—($n = 8$)
2. Group B: Measuring spatial skills/thinking on an environmental scale and in everyday life contexts [articles 9–13]—($n = 5$)

Table 4.1 Research studies reviewed

Reference	Research fields	Intended use of measure	Scale of spatial skills/thinking	Types of spatial skills/thinking	Context of spatial thinking	Response format of measure
GROUP A: Measuring spatial skills/thinking on a small scale without any specific context ($n = 8$)						
1. Cohen and Hegarty (2012)	Psychology	To measure ability to identify two-dimensional cross section of a three-dimensional geometric solid	Small	SV	Context free	Computer-based MCQ
2. Ramful et al. (2017)	Educational psychology	To measure ability to solve spatial representation tasks	Small	MR	Context free	MCQ
			Small	SO	Context free	
			Small	SV	Context free	
3. Burte et al. (2019)	Cognitive psychology	To assess STEM-specific spatial thinking	Small	SV	Context free	SAQ
4. Eilam and Alon (2019)	Science education	To develop an age-appropriate assessment and investigate the possible contribution of spatial ability to STEM learning	Small	SO	Context free	MCQ

(continued)

Table 4.1 (continued)

Reference	Research fields	Intended use of measure	Scale of spatial skills/thinking	Types of spatial skills/thinking	Context of spatial thinking	Response format of measure
5. Goldsmith et al. (2016)	Developmental psychology	To investigate the connections between visual-spatial thinking and geometric reasoning	Small	MR	Context free	Others (i.e., drawing task)
			Small	SV	Context free	
6. Gilligan et al. (2018)	Psychology	To establish developmental profile of scaling ability	Small	SV	Context free	Computer-based MCQ
7. Verdine et al. (2014)	Mathematics education	To investigate the relation between spatial and mathematical skill	Small	SV	Context free	Performance test
8. Clements-Stephens et al. (2013)	Cognitive psychology	To investigate relationships between perspective taking and social skills	Small	SO	Context free	Computer-based MCQ
Group B: Measuring spatial skills/thinking on an environmental scale and in everyday life contexts ($n = 5$)						
9. Nazareth et al. (2019)	Cognitive psychology	To compare wayfinding between different professions	Environmental	SO & Wayfinding	ST in everyday life	Computer-simulated MCQ
10. Lind et al. (2014)	Neurocognition	To investigate relationship between wayfinding and other cognitive abilities	Environmental	Wayfinding	ST in everyday life	Computer-simulated performance test

(continued)

Table 4.1 (continued)

Reference	Research fields	Intended use of measure	Scale of spatial skills/thinking	Types of spatial skills/thinking	Context of spatial thinking	Response format of measure
11. Descloux and Maurer (2018)	Neuropsychology/medicine	To assess ability to orient in a well-known environment	Environmental	Wayfinding	ST in everyday life	Performance test using mental maps
12. van der Ham et al. (2020)	Medicine/neuropsychology	To measure wayfinding ability to detect pathological aging	Environmental	Wayfinding	ST in everyday life	MCQ using video
13. Coutrot et al. (2019)	Neuroscience/Medicine	To quantify impairments in navigation performance for Alzheimer patients	Environmental	Wayfinding	ST in everyday life	Performance test using mobile app
Group C: Measuring spatial skills/thinking on a geographic scale and in a geoscience context (*n* = 3)						
14. Ishikawa (2013)	Geography education	To assess geospatial concepts and ability to think using the concepts	Geographic	SR	ST in geoscience	SAQ
15. Lee and Bednarz (2012)	Geography education	To develop a standardized test integrating geographical content knowledge and spatial skills	Small	SV	Context free	MCQ
			Geographic	SR	ST in geoscience	

(continued)

Table 4.1 (continued)

Reference	Research fields	Intended use of measure	Scale of spatial skills/thinking	Types of spatial skills/thinking	Context of spatial thinking	Response format of measure
			Geographic	SO	ST in geoscience	
16. Huynh and Sharpe (2013)	Geography education	To measure spatial thinking performance within a geographic context	Geographic	SO	ST in geoscience	MCQ, SAQ, others (i.e., drawing a diagram)
			Geographic	SR		
Group D: Spatial skills/thinking self-assessment in everyday life contexts (n = 3)						
17. Hegarty et al. (2010)	Psychology	To compare spatial abilities between different professions	Small	SV	ST in everyday life	LS
18. Erskine et al. (2015)	Business school	To assess geospatial reasoning ability	Environmental	Wayfinding	ST in everyday life	LS
			Environmental	SO		
			Geographic	SR		
19. Turgut (2015)	Educational psychology	To evaluate spatial abilities	Small	SV	ST in everyday life	LS
			Environmental	SO		
			Environmental	Wayfinding		

Note ST = Spatial thinking; SV = Spatial visualization; SO = Spatial orientation; MR = Mental rotation; SR = Spatial relations; MCQ = Multiple-choice question; SAQ = Short answer question; LS = Likert scale

3. Group C: Measuring spatial skills/thinking on a geographic scale and in a geoscience context [articles 14–16]—($n = 3$)
4. Group D: Spatial skills/thinking self-assessment in everyday life contexts [articles 17–19]—($n = 3$).

Overall, 19 different measurement instruments were identified within the included studies. Eight of them measure spatial skills/thinking on a small scale without any specific context (Group A). Five instruments focus on spatial skills/thinking on an environmental scale and in the context of everyday life (Group B). Three instruments aim to measure spatial skills/thinking on a geographic scale and in the context of geoscience (Group C). The remaining three measures are questionnaires evaluating individuals' spatial thinking in the context of everyday life (Group D). For each group, we discuss background of measurement (i.e., fields of research, intended use of measure), conceptual framework (i.e., scale and types of spatial skills/thinking tested, context of spatial thinking), and measurement mode (i.e., response format) with sample items.

4.4.1 Group A: Measuring Spatial Skills/Thinking on a Small Scale and Without Any Specific Context

Measurement tools in Group A focus on assessing spatial skills/thinking on a small scale and in a context-free environment. Small-scale spatial ability means the ability to mentally represent and transform two- and three-dimensional images, which can typically be understood from a single vantage point (Wang and Carr 2014). As a reminder, small scale spatial skills/thinking incorporates spatial visualization, spatial orientation, and mental rotation. The ability to generate, retain, and transform visual images is required to solve the items of these measurements (Lohman 1996). Typical test items tend to be articulated in a context-free environment. Since these tests aim to measure abilities rather than simply how well a person does on a specific test, it rarely includes any domain-specific knowledge or context-specific issues to influence an individual's performance on the test.

Papers 1–8 fall under this category. Most measurements (6 out of 8) are studied in the many sub-fields of psychology, including cognitive, educational, and developmental psychology. The other two papers are from science and mathematics education. The relationships between a high level of spatial ability and the possibility of success in STEM subjects has been among the most frequently studied topics by researchers in Group A ($n = 5$). The objectives of developing new spatial skills/thinking measurements are described in their works as follows: to measure ability to identify the two-dimensional cross section of a three-dimensional geometric solid, which is important in STEM fields (Cohen and Hegarty 2012); to measure spatial task representations that middle-school students may encounter in STEM subjects (Ramful et al. 2017); to develop assessments of spatial thinking in specific STEM disciplines (Burte et al. 2019); to investigate the possible contribution of spatial ability to STEM learning (Eilam and Alon 2019); and to investigate the relationship between spatial and mathematical skill (Verdine et al. 2014).

Multiple-choice questions are the most common response format and account for the largest proportion (5 out of 8) in this group. These questions task the examinee with identifying a pair of visual images or objects presented at different angles or shapes. For example, participants were asked to identify the two-dimensional shape that would result when the three-dimensional object is sectioned as indicated (Fig. 4.1A) or asked to find out which direction to look at the table to see the picture presented on the computer (Fig. 4.1B). Figures 4.1A and 4.1B were question items taken from the Cohen and Hegarty (2012) and Clements-Stephens et al. (2013) studies, respectively. There were some variations in the response format.

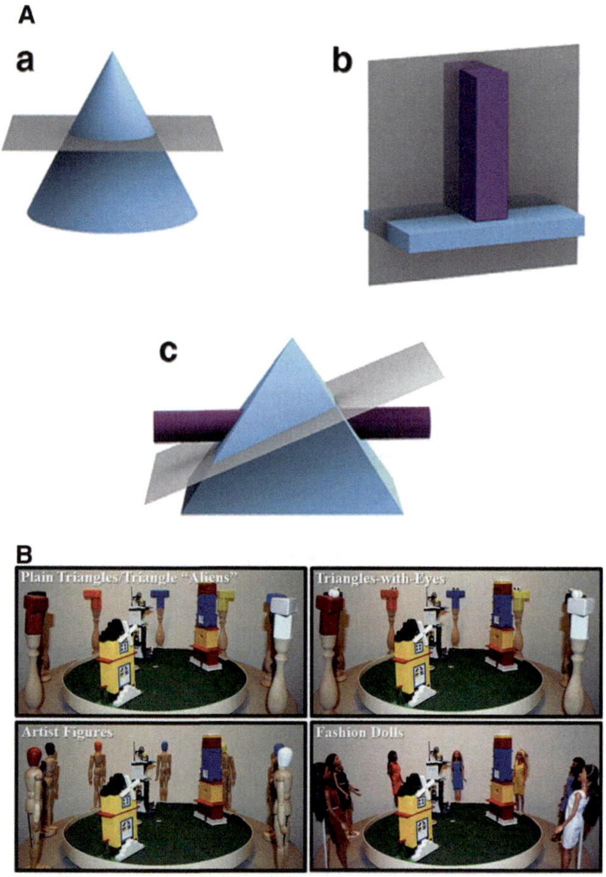

Fig. 4.1 Example items of spatial thinking assessment based on table-top scale and context free (*Source* Cohen and Hegarty 2012 [4.1A] [Reprinted from *Learning and Individual Differences*, Vol 22, Issue 6, CA Cohen and M Hegarty, Inferring cross sections of 3D objects: a new spatial thinking test, p. 869, © 2012, with permission from Elsevier]; Clements-Stephens et al. 2013 [4.1B])

Unlike most cases where participants need to find a pair of visual analogies, these two studies ask them to draw the appearance of the presented object when viewed from a different angle (Goldsmith et al. 2016) or to make blocks identical to those presented on the table (Verdine et al. 2014).

4.4.2 Group B: Measuring Spatial Skills/Thinking on an Environmental Scale and in Everyday Life Contexts

Measures in this group aim to assess spatial skills/thinking on an environmental scale and in everyday life. For the spatial tasks at the environmental scale, individuals carry out cognitive processing of spatial information 'in' the environment (Hegarty and Waller 2004). The type of spatial skills/thinking in Group B includes wayfinding, which is the ability to learn the layout of a new environment, retrace a route, or point to unseen locations in a known environment (Hegarty et al. 2006). As wayfinding is used by most people every day in a real-world setting, it has been assessed in the context of everyday life (Kitchin and Blades 2002).

Papers 9–13 fall under this category. The measurement of wayfinding is conducted in the field of neuroscience or psychology, including neuropsychology, neurocognition, and cognitive psychology. Out of the five papers in this group, the first two papers researched more traditional topics of psychology, such as the relationship between spatial navigation ability and other cognitive abilities, including IQ, language ability, episodic memory (Lind et al. 2014), and differences in spatial ability between professions, including geologists, psychologists, and Geographical Information Systems (GIS) students (Nazareth et al. 2019). The second group of papers ($n = 3$) offer new developments in objective measures in clinical contexts, including examining the impact of age and gender on spatial navigation ability in the context of a clinical setting (van der Ham et al. 2020), investigating the impact of age on the inability to orient in a well-known environment (Descloux and Maurer 2018), and quantifying Alzheimer patients' impairments in navigation performance (Coutrot et al. 2019).

Until recent years, the measurement of wayfinding had been limited due to the difficulty of utilising rigorous approaches that are standardized across participants in real-life settings (Malanchini et al. 2020). Recent technological advances in the field of virtual reality provide a novel powerful tool to study individuals' wayfinding. Compared with measurements taken in a real-world environment, the methods using a virtual environment allowed researchers to quickly, easily, and repeatedly record standardized information related to individuals' navigational skills (e.g., proportion of successful trials, walk time to reach goals, cumulative distance to the target) regardless of weather conditions (Lokka et al. 2018). Four papers in the selected works used either virtual reality or mobile apps to provide the test takers with real-world environments. Although the specific contexts are often different, the core task of the wayfinding assessment is similar; that is, participants are presented with an

Fig. 4.2 Example item of virtual wayfinding tasks represented in a virtual environment (*Source* Nazareth et al. 2019)

environmental-scale experimental environment consisting of landmarks and routes and are then asked to remember their location and spatial arrangement, to visualize buildings from different angles or perspectives, or to find a route to the destination based on route learning (Fig. 4.2). Figure 4.2 is a question item from the Nazareth et al. (2019) study.

4.4.3 Group C: Measuring Spatial Skills/Thinking on a Geographic Scale and in a Geoscience Context

Measurement tools in Group C focus on assessing spatial skills/thinking on a geographic scale and in a geoscience context. The geographic scale refers to the spatial range that cannot be seen by the human eye and requires an external representation like a map to be understood (Montello 1993). It is fundamentally two-dimensional and represented as a ratio between a distance on a map and its corresponding distance in the real world (Goodchild 2001; Mark and Freundschuh 1995). Types of spatial skills/thinking in Group C include 'spatial relations', defined by Golledge and Stimson (1997), or 'geospatial thinking', a term used to emphasize both its spatial nature and its geographic contents (Ishikawa 2013).

Papers 14–16 fall under this category. Research and measurement of geographic-scale spatial skills have been mostly conducted in 'spatial' disciplines such as geography and geosciences education. The objectives of developing spatial skills/thinking measurements are to assess them in a disciplinary context. For example, Lee and Bednarz (2012) aimed to develop a standardized test of spatial thinking that integrated geography content knowledge and spatial skills. The components of spatial relations as defined by Golledge and Stimson (1997) provided guidelines for developing test items. Ishikawa (2013) and Huynh and Sharpe (2013) identified the spatial thinking skills that are important and representative in the domain (e.g., geography) and then developed items to assess those skills (e.g., comprehending spatial correlations on a map).

Multiple-choice questions and performance tasks were used as measurements in Group C. These are often presented as simplified cartographic maps, aerial photos, and graphics as shown in Fig. 4.3, which displays questions taken from the Ishikawa (2013) and Lee and Bednarz (2012) studies, respectively. For example, the question in 4.3A in Fig. 4.3 assesses whether subjects can recognize, apply, and reason with the concept of spatial distribution. For this item, the subjects are allowed to use any method that they prefer (e.g., describing in words or using diagrams). Question 4.3B in Fig. 4.3 was developed to find out whether the subjects could comprehend and apply 'integration of geographic features represented as points, networks, and regions' (Golledge 2002) in a multiple-choice format.

4.4.4 Group D: Spatial Skills/Thinking Self-Assessment in Everyday Life Contexts

Group D is distinguished from other groups in terms of response format (i.e., Likert-style questionnaire) rather than the scale factor of spatial skills/thinking measurement. Likert-style ratings ask respondents to indicate agreement with a series of statements about their attitudes, self-rated abilities, and beliefs about spatial skills/thinking using a 5- or 7-point Likert scale, where 1 denotes 'strongly disagree' and 5 (or 7) denotes 'strongly agree'. Self-assessment questionnaires are believed to assess broader aspects of spatial thinking in both non-academic and academic areas (Hegarty et al. 2002), and there are a few examples of these available on the Internet (e.g., Golledge 2000, 2001). A typical question from these instruments might ask people to rate on a statement such as, 'When traveling, I take shortcuts as frequently as possible'. Although self-report tools are capable of assessing spatial thinking on both the small and large (environmental and geographic) scale (Hegarty et al. 2002) and are useful in assessing individuals' spatial behaviors in everyday life, researchers recognized that the tools do not directly measure a cognitive ability but rather a perception of ability.

Papers 17–19 fall under this category. Measurements of Group D are used in various research areas including psychology and business schools. More specifically, these papers addressed the spatial ability of individuals unfamiliar with cartographic or geographic terms (Erskine et al. 2015) to compare the spatial ability of experts in various fields including geographers, engineers, and humanists (Hegarty et al. 2010) or to determine the level of spatial thinking ability of undergraduate students (Turgut 2015).

The measurement in Group D used a Likert-style questionnaire covering all scale ranges from small to environmental and geographic and mostly in everyday life contexts. Figure 4.4 shows the question items designed to assess mental rotation,

A

Suppose that the three maps below show the locations of traffic accidents that happened in three different regions (within an area of 1 km by 1 km) for the past one month. One dot represents one accident.

Q1. Explain the characteristics of the three maps in as much detail as possible, by comparing them with each other.

Q2. What do you think are the possible reasons underlying these three maps of traffic accidents?

B

Real-world objects can be represented explicitly by point, line (arc), and area (polygon). Classify the following spatial data.

Q. Locations of weather stations in Washington County.

A. Lines
B. Area
C. Points and Lines
D. Points and Area

Fig. 4.3 Example items of spatial thinking assessments based on geographic scale and the context of geoscience (*Source* Ishikawa 2013 [4.3A] [Reprinted from *The Professional Geographer*, Vol 65, Issue 4, T Ishikawa, Geospatial thinking and spatial ability: an empirical examination of knowledge and reasoning in geographical science, © 2013, reprinted by permission of Taylor & Francis Ltd.]; Lee and Bednarz 2012 [4.3B] [Reprinted from *Journal of Geography*, Vol 111, Issue 1, J Lee and R Bednarz, Components of spatial thinking: evidence from a spatial thinking ability test, © 2012, reprinted by permission of Taylor & Francis Ltd.])

1. I can imagine the rotated versions of three-dimensional objects mentally.
2. I can fold and create new shapes from a square piece of paper in my mind.
3. I feel that I can easily orientate myself in a new place.
4. I can find a shorter route to a location if I have been there before.

Fig. 4.4 Example items of spatial thinking assessment using a Likert-scale questionnaire in everyday life (*Source* Erskine et al. [2015] [Reprinted from *International Journal of Human–Computer Interaction*, Vol 31, Issue 6, MA Erskine et al., Geospatial reasoning ability: definition, measurement and validation, © 2015, reprinted by permission of Taylor & Francis Ltd.]; Turgut [2015])

spatial visualization, spatial orientation, and wayfinding ability respectively. These items were from studies by Erskine et al. (2015) and Turgut (2015).

4.5 Conclusion

This chapter utilized a modified systematic review protocol to investigate the body of research related to assessment of spatial thinking/skills since the 2006 publication of *Learning to Think Spatially*. Nineteen papers were identified which aimed to measure spatial skills/thinking. Data related to these studies' conceptual frameworks and methods were analyzed according to scale (i.e., small, environmental, geographical), contexts of spatial thinking (i.e., context free, context of science, context at work, context of everyday life), and response format (i.e., assessing ability vs. assessing perception). The four groups (Group A–D) identified are represented in Fig. 4.5.

Group A represents a research tradition in the field of cognitive psychology that has measured small-scale spatial ability in a context-free manner. This is the group that is actively conducting most research and in particular focuses on identifying the link between spatial thinking and STEM, which has recently attracted attention socially and educationally. Group B research is also closely related to the tradition of cognitive psychology. Group B focuses on spatial thinking skills on an environmental scale that can be experienced through body movement (e.g., wayfinding ability). The advancement of technologies such as virtual reality has helped overcome the limitations of measuring wayfinding abilities in real-world settings, and, recently, collaboration with applied fields such as medicine and health sciences are notable. While Group A and Group B focus on general spatial skills/thinking, Group C emphasizes and attempts to measure domain-specific spatial thinking. Group C aims to respond to the claims and demands of *Learning to Think Spatially*, which argues that since we cannot teach or learn spatial thinking without a specific context, it is difficult to evaluate such skills without a context. In Group C, three notable studies have been published, each of which has developed new assessment methods based on different sets of spatial concepts on a mostly geographical scale. The last is Group D, which is different from the three groups mentioned earlier in terms of response format. While

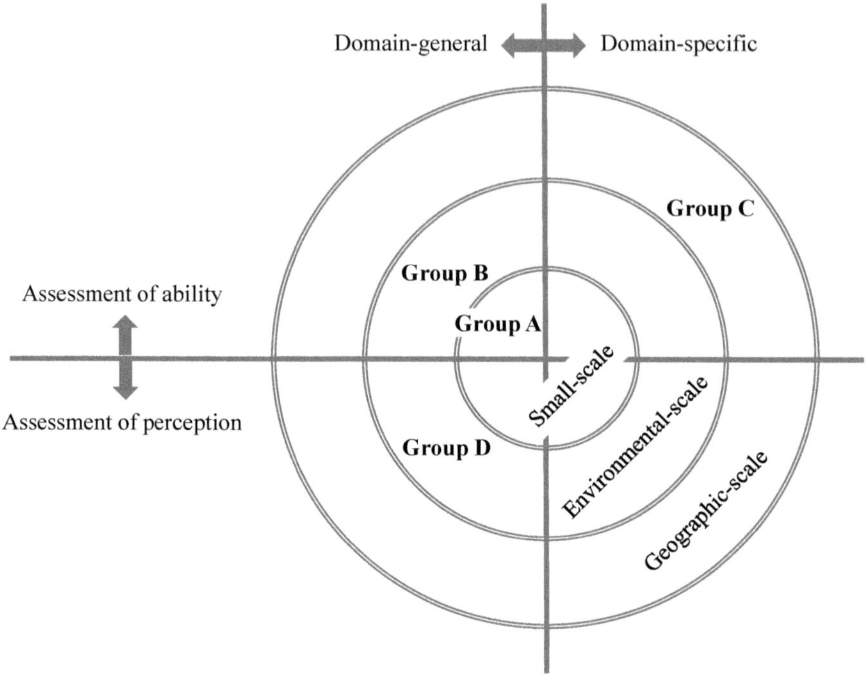

Fig. 4.5 Current knowledge about assessing spatial skills/thinking (ST)

the first three groups tried to directly measure an individual's spatial skills/thinking, Group D measures spatial thinking using self-report instruments. However, one's real skills and one's perceptions of skills do not always coincide. Since Group D is composed of Likert scale questionnaires, it is not only easy to develop but also can measure spatial thinking skills of various scales.

Learning to Think Spatially has had an impact on the field of research in spatial thinking and its measurement/assessment, but as this study shows, research in this space is still limited with existing problems, such as differences in concepts of spatial thinking across disciplines and a lack of reliable measurement tools persisting. For this reason, it is not easy to share research findings between disciplines. At the present time, we need to synthesize and understand, as a whole, the research being conducted in each area under the name of spatial thinking measurement/assessment rather than expecting spatial thinking measurement tools to satisfy the prerequisites and expectations of all research groups. The progress of research in interdisciplinary fields such as spatial thinking is only possible through understanding and comparison across other fields, and in this respect, the present study can contribute to the further development of spatial thinking measurement research. As Fig. 4.5 shows, there is an absence of research about people's perceptions of their spatial skills/thinking across all scales in specific domains. This could be a future direction for progressing research in this field.

References

Alias M, Black T, Gray D (2002) Effect of instructions on spatial visualisation ability in civil engineering students. Int Educ J 3(1):1–12

Bednarz R, Lee J (2019) What improves spatial thinking? Evidence from the spatial thinking abilities test. Int Res Geogr Environ Educ 28(3):1–19

Bednarz SW, Kemp K (2011) Understanding and nurturing spatial literacy. Procedia Soc Behav Sci 21:18–23

Burte H, Gardony AL, Hutton A, Taylor HA (2019) Make-A-Dice Test: assessing the intersection of mathematical and spatial thinking. Behav Res Meth 51(2):602–638. https://doi.org/10.3758/s13428-018-01192-z

Carroll JB (1993) Human cognitive abilities: a survey of factor-analytic studies. Cambridge University Press, Cambridge

Clements-Stephens AM, Vasiljevic K, Murray AJ, Shelton AL (2013) The role of potential agents in making spatial perspective taking social. Front Hum Neurosci 7:497. https://doi.org/10.3389/fnhum.2013.00497

Cohen CA, Hegarty M (2012) Inferring cross sections of 3D objects: a new spatial thinking test. Learn Indiv Differences 22:868–874

Coutrot A, Schmidt S, Coutrot L, Pittman J, Hong L, Wiener JM, Hölscher C, Dalton RC, Hornberger M, Spiers HJ (2019) Virtual navigation tested on a mobile app is predictive of real-world wayfinding navigation performance. PLoS One 14(3):e0213272. https://doi.org/10.1371/journal.pone.0213272

Descloux V, Maurer R (2018) Cognitive map recall test: a new specific test to assess topographical disorientation. Appl Neuropsychol Adult 25(2):91–109. https://doi.org/10.1080/23279095.2016.1247094

Eilam B, Alon U (2019) Children's object structure perspective-taking: training and assessment. Int J Sci Math Educ 17(8):1541–1562. https://doi.org/10.1007/s10763-018-9934-7

Ekstrom RB, French JW, Harman HH, Dermen D (1976) Manual for kit of factor-referenced cognitive tests. Princeton, Educational Testing Services

Eliot J, Czarnolewski MY (2007) Development of an everyday spatial behavioral questionnaire. J Gen Psychol 134(3):361–381

Erskine MA, Gregg DG, Karimi J, Scott JE (2015) Geospatial reasoning ability: definition, measurement and validation. Int J Hum Comput Interact 31(6):402–412. https://doi.org/10.1080/10447318.2015.1034551

Gersmehl PJ, Gersmehl CA (2006) Wanted: a concise list of neurologically defensible and assessable spatial-thinking skills. Res Geogr Educ 8:5–38

Gilligan K, Hodgkiss A, Thomas MSC, Farran EK (2018) The use of discrimination scaling tasks: a novel perspective on the development of spatial scaling in children. Cognit Dev 47:133–145. https://doi.org/10.1016/j.cogdev.2018.04.001

Goldsmith LT, Hoyle C, Hetland L, Winner E (2016) Visual-spatial thinking in geometry and the visual arts. Psychol Aesthet Creativ Arts 10(1):56–71

Golledge RG (1993) Geographical perspectives on spatial cognition. In: Gärling T, Golledge R (eds) Behavior and environment: psychological and geographical approaches. Elsevier Science Publishers, Amsterdam, pp 16–45

Golledge RG (2000) Geography and everyday life. Directions Magazine. https://www.directionsmag.com/article/4035. Accessed 10 Jan 2020

Golledge RG (2001) Geography and everyday life (again!). Directions Magazine. https://www.directionsmag.com/article/4021. Accessed 10 Jan 2020

Golledge RG (2002) The nature of geographic knowledge. Ann Assoc Am Geogr 92(1):1–14

Golledge RG (2003) Thinking spatially. Directions Magazine. https://www.directionsmag.com/article/3803. Accessed 10 Jan 2020

Golledge RG, Stimson RJ (1997) Spatial behavior: a geographic perspective. The Guilford Press, New York

Goodchild MF (2001) Models of scale and scales of modelling. In: Tate NJ, Atkinson PM (eds) Modelling scale in geographical information. Wiley, Chichester, pp 3–10

Hegarty M, Crookes RD, Dara-Abrams D, Shipley TF (2010) Do all science disciplines rely on spatial abilities? Preliminary evidence from self-report questionnaires. In: Proceedings of the International Conference on Spatial Cognition Spatial Cognition: spatial cognition VII, pp 85–94

Hegarty M, Montello DR, Richardson AE, Ishikawa T, Lovelace K (2006) Spatial abilities at different scales: individual differences in aptitude-test performance and spatial-layout learning. Intelligence 34(2):151–176

Hegarty M, Richardson AE, Montello DR, Lovelace K, Subbiah I (2002) Development of a self-report measure of environmental spatial ability. Intelligence 30(5):425–447

Hegarty M, Waller D (2004) A dissociation between mental rotation and perspective-taking spatial abilities. Intelligence 32(2):175–191

Hespanha SR, Goodchild F, Janelle DG (2009) Spatial thinking and technologies in the undergraduate social science classroom. J Geogr High Educ 33(1):17–27. https://doi.org/10.1080/03098260903033998

Huynh NT, Sharpe B (2013) An assessment instrument to measure geospatial thinking expertise. J Geogr 112(1):3–17

Ishikawa T (2013) Geospatial thinking and spatial ability: an empirical examination of knowledge and reasoning in geographical science. Prof Geogr 65(4):636–646. https://doi.org/10.1080/00330124.2012.724350

Jones S, Burnett G (2008) Spatial ability and learning to program. Hum Tech 4(1):47–61

Kitchin R, Blades M (2002) The cognition of geographic space. IB Tauris, London

Lee J (2005) The effect of GIS learning on spatial ability. Dissertation, Texas A&M University

Lee J, Bednarz R (2009) Effect of GIS learning on spatial thinking. J Geogr High Educ 33(2):183–198

Lee J, Bednarz R (2012) Components of spatial thinking: evidence from a spatial thinking ability test. J Geogr 111(1):15–26. https://doi.org/10.1080/00221341.2011.583262

Liben LS (1981) Copying and reproducing pictures in relation to subjects' operative levels. Dev Psychol 17(3):357–365. https://doi.org/10.1037/0012-1649.17.3.357

Lind SE, Bowler DM, Raber J (2014) Spatial navigation, episodic memory, episodic future thinking, and theory of mind in children with autism spectrum disorder: evidence for impairments in mental simulation? Front Psychol 5:1411. https://doi.org/10.3389/fpsyg.2014.01411

Linn MC, Petersen AC (1985) Emergence and characterization of sex differences in spatial ability: a meta-analysis. Child Dev 56(6):1479–1498

Lohman DF (1996) Spatial ability and g. In: Dennis I, Tapsfield P (eds) Human abilities: their nature and measurement. Lawrence Erlbaum Associates, Mahwah, pp 97–116

Lokka IE, Çöltekin A, Wiener J, Fabrikant SI, Röcke C (2018) Virtual environments as memory training devices in navigational tasks for older adults. Sci Rep 8(10809):1–15

Malanchini M, Rimfeld K, Wang Z, Petrill SA, Tucker-Drob EM, Plomin R, Kovas Y (2020) Genetic factors underlie the association between anxiety, attitudes and performance in mathematics. Translational Psychiatr 10:12. https://doi.org/10.1038/s41398-020-0711-3

Mark DM, Freundschuh SM (1995) Spatial concepts and cognitive models for geographic information use. In: Nyerges TL, Mark DM, Laurini R, Egenhofer MJ (eds) Cognitive aspects of human–computer interaction for geographic information systems. NATO ASI Series (Series D: Behavioural and Social Sciences) vol 83. Springer, Dordrecht. https://doi.org/10.1007/978-94-011-0103-5_3

Marunic G, Glazar V (2013) Spatial ability through engineering graphics education. Int J Tech Des Educ 23(3):703–715

McGee MG (1979) Human spatial abilities: psychometric studies and environmental, genetic, hormonal, and neurological influences. Psychol Bull 86(5):889–918. https://doi.org/10.1037/0033-2909.86.5.889

Montello D (1993) Scale and multiple psychologies of space. In: Frank A, Campari I (eds) Spatial information theory: a theoretical basis for GIS. Lecture notes in computer science vol 716. Springer-Verlag, Berlin, pp 312–321

National Research Council (2006) Learning to think spatially. The National Academies Press, Washington, DC
Nazareth A, Newcombe NS, Shipley TF, Velazquez M, Weisberg SM (2019) Beyond small-scale spatial skills: navigation skills and geoscience education. Cognit Res 4:17. https://doi.org/10.1186/s41235-019-0167-2
Ramful A, Lowrie T, Logan T (2017) Measurement of spatial ability: construction and validation of the spatial reasoning instrument for middle school students. J Psychoeduc Assess 35(7):709–727
Self CM, Golledge RG (1994) Sex-related differences in spatial ability: what every geography educator should know. J Geogr 93(5):234–243
Shepard S, Metzler D (1988) Mental rotation: effects of dimensionality of objects and type of task. J Exp Psychol Hum Percept Perform 14(1):3–11
Small M, Morton M (1983) Research in college science teaching: spatial visualization training improves performance in organic chemistry. J Coll Sci Teach 13(1):41–43
Sutton K, Allen R (2011) Assessing and improving spatial ability for design-based disciplines utilising online systems. Australian Learning and Teaching Council, Sydney
Turgut M (2015) Development of the spatial ability self-report scale (SASRS): reliability and validity studies. Qual Quant Int J Meth 49(5):1997–2014. https://doi.org/10.1007/s11135-014-0086-8
Vandenberg SG, Kuse AR (1978) Mental rotations, a group test of three-dimensional spatial visualization. Percept Mot Skills 47(2):599–604. https://doi.org/10.2466/pms.1978.47.2.599
van der Ham IJM, van der Kuil MNA, Claessen MHG (2020) Quality of self-reported cognition: effects of age and gender on spatial navigation self-reports. Aging Ment Health. https://doi.org/10.1080/13607863.2020.1742658
Verdine BN, Golinkoff RM, Hirsh-Pasek K, Newcombe NS, Filipowicz AT, Chang A (2014) Deconstructing building blocks: preschoolers' spatial assembly performance relates to early mathematical skills. Child Dev 85(3):1062–1076
Voyer D, Voyer S, Bryden MP (1995) Magnitude of sex differences in spatial abilities: a meta-analysis and consideration of critical variables. Psychol Bull 117(2):250–270. https://doi.org/10.1037/0033-2909.117.2.250
Wai J, Lubinski D, Benbow CP (2009) Spatial ability for STEM domains: aligning over fifty years of cumulative psychological knowledge solidifies its importance. J Educ Psychol 101(4):817–835
Wakabayashi Y, Ishikawa T (2011) Spatial thinking in geographic information science: a review of past studies and prospects for the future. Procedia Soc Behav Sci 21:304–313
Wang L, Carr M (2014) Working memory and strategy use contribute to gender differences in spatial ability. Educ Psychol 49(4):261–282
Wang L, Carr M (2020) Understanding the relationship between spatial ability and mathematics achievement through the lens of information processing theory. N Am J Psychol 22(2):173–192
Xie F, Zhang L, Chen X (2020) Is spatial ability related to mathematical ability: a meta-analysis. Educ Psychol Rev 32(1):113–155

Jongwon Lee is a Professor of Social Studies Education (geography education major) and is currently serving as Associate Dean for College of Education at Ewha Womans University, Seoul. His research interests include inquiry-based learning, virtual fieldwork with VR, and spatial thinking abilities. He has served as Secretary of the Commission on Geographical Education (CGE) of the International Geographical Union (IGU) since August 2016.

Injeong Jo is an Associate Professor of Geography and Environmental Studies at Texas State University. Her research focuses on geography and spatial thinking education, geospatial technologies for teaching and learning, as well as teacher education. She teaches courses in geographic education for both undergraduate and graduate students, including Fundamental Themes in Geography, Spatial Thinking in Education, and Technology in Geographic Education. She currently serves as the Faculty Coordinator of the Online Master of Applied Geography in Geographic Education Program.

Part III
Assessment Related to the Signature Pedagogies of Inquiry and Fieldwork

Chapter 5
Integrating Assessment Effectively into International Fieldwork: A Case Study Using Student-Led Teaching and Learning

Alan Marvell and David Simm

Abstract Assessing international fieldwork provides a series of challenges and opportunities that frames the student experience. In this case study students are presented with a task to deliver an assessed presentation and an activity based on a research topic of their choice to their peers in situ. Students are assessed and gain formative and summative feedback through multiple stages of the fieldtrip module from an initial group project proposal, a group presentation, an individual fieldwork diary, a reflective essay on their student-led experience, and an essay based on an overarching research theme. The chapter reveals that assessment and feedback are inextricably linked with one stage of the process informing another. Direct experiences of being on fieldwork stimulates the affective domain, with resulting emotional responses and behavior influencing performance (in assessment) and participation in students' communities of learning (including group dynamics). This highlights the importance and role of (self-)reflection in assessment. Further consideration is made to adapt the approach to assessing international fieldwork at other levels and scales of geographical education.

Keywords Assessment · Fieldwork · Student-led teaching · Geography · Fieldwork diary · Affective domain

5.1 Introduction

Fieldwork has always been at the heart of geography and is considered its signature pedagogy (Kim 2020). Through enquiry, students are able to engage in experiential and discursive learning environments where the very nature of the discipline and its pedagogies are interwoven (Hill et al. 2018). Fieldwork is an effective mode

A. Marvell (✉)
University of Gloucestershire, Oxstalls Campus, Gloucester, UK
e-mail: amarvell@glos.ac.uk

D. Simm
Bath Spa University, Newton Park, Bath, UK

T. Bourke et al. (eds.), *Assessment in Geographical Education: An International Perspective*, Key Challenges in Geography,
https://doi.org/10.1007/978-3-030-95139-9_5

of learning in geography, providing experience of the real world and developing a range of transferable skills (Fuller et al. 2006) that are of a direct benefit to students in terms of future studies, life skills, and employability (Arrowsmith et al. 2011; Peasland et al. 2019). Skills associated with fieldwork include critical thinking, problem solving, communication, and leadership as well as collaboration, adaptability, and curiosity (Ito and Igano 2021), and interviewing and negotiation skills (Hindle 2000). Demonstrating and applying knowledge and understanding of theory is also integral to fieldwork (Fuller et al. 2006).

Student-centered approaches to field courses allow students to demonstrate a wide range of skills as well as being empowered to make decisions (Holton 2017). The focus of the fieldwork can take a variety of forms (Kent et al. 1997) such as: a descriptive-explanatory focus that is based on observational and inferential methods, designed to offer a factual introduction to the subject and can also include stop-and-look or guided trails (Coe and Smyth 2010); an analytical-predictive focus based on participatory and deductive methods through problem-based learning and group work (Pawson et al. 2006) and may also involve specific research methods training and the use of specialist equipment; and, finally, a learner-practitioner focus that can involve students as participant observers.

International fieldwork immerses students within an unfamiliar environment or location which challenges their preconceptions and enhances an affective response through emotions and feelings, which in turn provides a deeper learning experience (Ito and Igano 2021). As such, students are expected to apply their knowledge and reflect to better understand and explain geographical concepts and events. Kolb's (1984) experiential learning theory is a key part of reflective learning as students move through the successive stages of concrete experience (new experience or situation), reflective observation (reflection), abstract conceptualization (new idea or understanding) to active experimentation (application or demonstration). Selecting an international fieldwork destination situates students in an unfamiliar setting thus providing greater opportunities to create an immersive fieldwork experience and a potential to demonstrate higher-order cognitive skills (Simm and Marvell 2015). It can also provide a transformative learning experience (Mezirow 2000) by encouraging new situational encounters and opportunities to co-create knowledge and, through increasing cultural and ethical awareness, gain a greater sense of place. Fieldwork is so important that assessment design should acknowledge these influences through specific marking criteria and affirm them through grading and feedback.

5.2 Fieldwork, Students and Assessment

When assessing fieldwork it is important to convey what, how, when, and why assessments are required. Through a process of constructive alignment module outcomes, intended learning outcomes, and assessment criteria should coincide (Trigwell and Prosser 2014). The approach is based on the idea that students actively construct knowledge through learning. It is not something that is simply transmitted from the tutor but something that students explore and learn for themselves, and as such teaching and the wider fieldwork experience becomes a catalyst for learning (Biggs 2003). The tutor's role is therefore to facilitate the learning process by providing the structure and framework of the fieldwork situation or environment in which learning takes place. University tutors who devise student-centered assessment are more likely to view intended learning outcomes for students in more holistic terms and as having a greater integral role within curriculum design (Trigwell and Prosser 2014). But it is useful to remember that for the tutor the assessment is the end point of the learning process whereas most students will see the assessment as the beginning (Biggs 2003). Most tutors foreground their modules or courses with an assessment briefing to clearly communicate the learning journey to their students.

The tutor needs to be clear on the purpose of assessment. Assessment provides feedback to tutors and students to support future learning (formative) but can also indicate a level of attainment (summative) (Taras 2005). However, assessment is not just a method of measuring attainment; it should also be used to promote learning and to reflect upon experiences of participating in fieldwork. Creating assessment opportunities is not just about requiring students to repeat content that has been taught; it provides opportunities to create new knowledge, understanding, and experiences. As such, knowledge can be fixed and external to the learner and/or can be co-constructed by the learner whilst immersed during fieldwork (Hargreaves 2005). During the learning process students will desire feedback. They need to know how they are progressing but also establish what will improve their summative performance. However, when feedback is offered, students may not recognize it as such, or even when they are presented with feedback they may be unsure how best to use it to their advantage (Francis et al. 2019). It is therefore important that feedback is progressive and at best there are multiple opportunities for students to receive critical yet constructive comments (Nicholson 2011). Feedback may be given by the tutor, but it can also be given by the peer-group.

The involvement of students as learners is important through peer-group discussions and whole-class discussions to create a community or communities of learners, 'where the social learning of students … become[s] more salient and effective' (Black and Wiliam 2012, p. 26). By working in small groups on their research projects, students are able to share ideas and learn from each other. Discussion and debate within the group provides informal formative feedback on ideas and suggestions. Building group cohesion, shared practice, and identity is necessary to create a learning community as a 'group formed through mutual engagement, joint enterprise, and a shared repertoire' (Cuddapah and Clayton 2011, p. 64). As such, group fieldwork can create communities of practice (Wenger 2001) where students can learn better together (Tinto 2000) and result in the group achieving goals that they would not be able to achieve as individuals. However, some groups may not be cohesive and some members are less able to demonstrate 'self-regulation' through their individual capacities, be they cognitive, social, or behavioral (Dann 2014). Encouraging effective self-regulation is made possible by moving 'to a more social and situated view of learning and a shift from a focus on the individual as learner to learning as participation in the social world' (Graven and Lerman 2003, p. 186). As such the emphasis shifts from the egocentric to the social by engaging with others and through the use of language and social interactions (Vygotsky 1978 in Dann 2014). Students have the potential to engage more fully with their peers and with their immediate social and physical environment. Learning from experience is enhanced by fieldwork as the use of the senses and emotions are heightened. These can be explored through personal reflection where students record and critically analyze what they have experienced, leading to greater powers of awareness and observation (Moon 1999): 'Reflective learning is essentially the process of considering what we know or have learnt in order to generate new knowledge' (Dummer et al. 2008, p. 460).

Being immersed in the field is an emotional experience and as such this helps to define our understanding of those places and the processes that occur (Widdowfield 2000). It is important to design a fieldtrip and assessment that acknowledges the role of emotions. Non-representational theorists highlight the importance of performances, activities, and embodied experiences that can be experienced and expressed through emotion (Thrift 2007; McCormack 2008). Students are present in situ and not invisible to the activities of the place, they are intrinsically part of the activities. It is this relationship between being present and acknowledging that the thoughts and feelings experienced by students can also affect their sense of place (Marvell and Simm 2018).

Students learn through experiences and performance during fieldwork through the affective domain as they receive and respond to phenomena, and by internalizing their responses they begin to make value judgements and begin to organize and internalize their values that transform their understanding and behaviors (Krathwohl et al. 1964; Krathwohl 2002). A self-reflective diary or notebook provides opportunities to reveal the depth of student understanding and to communicate their thoughts and feelings (McGuinness and Simm 2005). This promotes deeper learning where complex issues are explored and reflected upon. The self-reflective diary is distinct from a traditional geography fieldwork notebook as the processes, thoughts, feelings, and emotional responses gain significance over the reporting of geographical data (Punch 2012). It enables students to make sense of being in place, offers a record of the experiences, thoughts, and ideas for subsequent assessment, and helps students to reflect and work through their experiences. The self-reflective diary, which is usually handwritten in the field, can be assessed formatively or as a summative assessment.

5.3 Fieldtrip Pedagogy

The Advanced Geographical Fieldtrip module is offered to final year undergraduate geography and tourism management students at Bath Spa University, UK. The main focus of the module is to encourage students to experience different landscapes and environments and enable them to make the connection between people and place. They are not in situ to just observe; they will also offer a valid explanation of what they see and experience and identify interconnections between geographical processes and places.

The aims and learning outcomes are to:

- Attain a geographical sense of place
- Actively be involved in logistical planning
- Work in groups to deliver a student-led field presentation and field activity
- Conduct independent and advanced research
- Demonstrate teamwork and project management skills
- Have the confidence and ability to cope with unfamiliar environments
- Demonstrate a critical self-appraisal of field experience and performance.

The module is delivered through a series of lectures and workshops that foreground the five-day fieldtrip to the city of Barcelona, Spain. Students work in groups and identify a topic to research before the fieldtrip. Their task is to deliver a 90-minute presentation with an activity in-the-field. The students organize themselves in terms of planning and preparation with support from a tutor. The workshops help to introduce students to the study area and to facilitate a discussion in groups regarding potential student-led research projects. Students are required to submit a pre-placement report outlining their project, proposed itinerary, and activity. The report is assessed on a pass/fail basis and provides formative feedback from tutors. This form of assessment is important as it allows students to shape and refine their original ideas, providing support yet challenging the students to consider alternatives. The concept of formative feedback was first proposed by Scriven (1967) to help facilitate on-going curriculum improvement (Bennett 2011). Bloom (1969) then used a similar approach but with a focus on student feedback, providing 'feedback and correctives at each stage in the teaching-learning process' (p. 48). Although Bloom (1969) initially used a series of brief tests, these were independent of the grading process and designed to guide students' learning.

By submitting a pre-fieldwork project report, students engage in discussion and evaluation of what is desirable and possible. They learn as a group, sharing ideas and exploring alternatives. Through dialogue with their peers they deliver and receive feedback and are informally assessed by their group members in terms of the potential success of the fieldwork project. Tutors supply formative feedback on the group project proposals which is used by students to improve on their project design and in most cases to narrow their intended focus as most initial proposals tend to be over ambitious.

The five-day fieldtrip begins with a staff-led familiarization tour and orientation activities followed by time given to students to undertake reconnaissance and to explore their research site. This is then followed by two days of student-led tours. The presentations and activities are mostly pre-planned and pre-scripted with handouts, but time at their research sites allows them to make final alterations and gives them a chance to fully interweave their sites within the presentation. The field activity is to encourage the student audience to participate fully in the presentation, not just writing field notes but also making observations and collecting data with the purpose of increased engagement and thereby promoting interaction and debate. Previous versions of the fieldtrip did not contain the activity and as such the audience, and to some extent the presentations, did not always fully engage with their environs whilst being immersed in situ (Marvell 2008).

Each morning students are given a self-reflective question which they are expected to consider and clarify in their field notebook during the evening. At the end of the fieldtrip, an overarching self-reflective theme is announced, and this draws upon their experiences and self-reflective entries during the field course, which they write up post-fieldtrip. They are encouraged to use pedagogic literature to help inform their reflection. The fieldtrip is then encapsulated in a final individual essay on the transformation of Barcelona to apply their experiences and knowledge of the city to a more overarching conceptualization of urban development.

The assessment for the fieldwork module can be summarized:

- Group pre-fieldtrip project proposal—Pass/Fail
- Group field presentation and activity—40%
- Field notebook and self-reflective essay—20%
- Essay: 'Transformation of Barcelona'—40%

5.4 Characteristics of Fieldwork Pedagogy

The underlying principles behind our field teaching strategies are to provide students with an opportunity to demonstrate their geographical knowledge through research of an unfamiliar location. By encountering liminal or borderland learning spaces that are unfamiliar in terms of environment/landscape and society/culture, students are challenged to understand and make sense of their encounters (Hill et al. 2016). By giving students responsibility to determine locations, research topics, and field activities, choosing their own research projects, identifying locations to visit, and essentially running part of the fieldtrip empowers them to claim greater ownership. By handing over some control of power for teaching and learning, students begin to make their own decisions and take responsibility for the teaching of others through student-led teaching. Empowerment allows students to develop and mature academically as they begin to develop and demonstrate aspects of 'scholarship' (Marvell et al. 2013).

Placing students at the center of learning challenges power relations and leads to transformative learning through experience and reflection. Reflection is important to help students make sense of what they have experienced. The changing power relations between staff and students can be unsettling for both groups. Giving power is not necessarily a natural action of the tutor whereas students may feel uncomfortable being given additional responsibilities for teaching and learning. As such, students start to become the 'authority' of their research topics and study locations, potentially taking tutors to new locations and revealing new knowledge. This approach enables co-partnership and co-production between students and staff (Hill et al. 2018). Students demonstrate responsibility not only for themselves and their own group but also for others in terms of logistics and learning. By encouraging students to gain a sense of ownership and self-direction, tutors are requiring students to demonstrate aspects of self-authorship (Magolda 2014). This requires skills of critical analysis, evaluation, self-reflection, and being able to empathize with diverse perspectives (Hill et al. 2016).

Sharing teaching with their peers extends the co-production of knowledge beyond the research group to a community of learners on the fieldtrip. They therefore take possession of their own learning and that of others. Through these fieldwork pedagogies students achieve transformative learning through being challenged by new experiences and knowledge as well as personal realization through (self-)reflection (Simm and Marvell 2015). However, with attempts to innovate and try something different with our students, we need to be more acutely aware of what they are feeling and how they are responding. As tutors, our approaches to teaching and learning need to work with students' emotional intelligence rather than against it. To begin with, tutors need to consider the feelings and emotions of students when engaged in fieldwork (Marvell and Simm 2018) as tutors have a duty of care to ensure the well-being of their students. Students who demonstrate an enhanced sense of well-being potentially have a stronger readiness to learn and have a greater sense of confidence, potentially leading to greater academic success (Scheyvens et al. 2003). Tutors need to be mindful that some students may have unproductive and negative experiences, and as such they may need to be guided through this by receiving feedback and support from the tutor. Thus, this particular fieldtrip experience and form of assessment can be described as 'disruptive', having the potential to place students outside their comfort zone in areas that are regarded as a 'porous boundary between the formal curriculum, co-curriculum and informal learning' (Higgitt 2014, p. 2). Within these borderland spaces of learning, students encounter the perimeters of familiarity through unfamiliar locations, environments, and cultures (Hill et al. 2016).

5.5 Learning and Teaching Strategies

It is not only about where we go with students, but also about what we do with them. Allowing students to challenge and be challenged and to achieve something that they would not be ordinarily able to do is part of learning. Students will have emotional responses to what they experience during fieldwork and also in reaction to the assessment tasks set by tutors (Marvell and Simm 2018). Emotions are most acute during times of stress and enjoyment which can be exacerbated by being in an unfamiliar place and with peers. As a result, students express a multitude of emotions (e.g., feeling challenged, anxious, excited, apprehensive, etc.). Tutors need to recognize that international fieldwork is emotionally challenging and tiring and as such has an effect on the delivery and engagement of the assessed presentations in situ. Such challenges test the character of each student and each may react and respond differently. As such these emotional encounters need to be managed carefully and taken into consideration when devising assessments and providing feedback.

When emotions are heightened, tutors need to be sensitive to the way in which critical formative feedback is offered to students. There is a need to become an emotionally intelligent teacher by offering acknowledgement, listening, engaging in dialogue, and responding appropriately (Mortiboys 2012). On one occasion a tutor encountered an adverse reaction to critical feedback after an assessed presentation on coastal management in Barcelona. The group of students selected a good location to deliver their presentation but overlooked a key feature, an offshore breakwater, that was integral to the story at that locality. At the end of the presentation during a request by the group for questions, the tutor asked for clarification on the feature and its importance. The purpose of the questions was to allow students to elaborate and explore content that was not fully covered during the presentation. As a summative assessment, the group could benefit from an improved grade if the group responded satisfactorily. However, during a heightened emotional state and the relief of finishing the presentation, the group reacted negatively to the question which was perceived as 'public humiliation' and 'unfair criticism'. This emotional reaction was caused by the pressures of the summative presentation and the perceived unfair treatment in front of their peers. As tutors we need to be mindful of what we say and do and employ a 'gentle criticism', as appropriate (Hill et al. 2021). Thus, emotions are typically heightened and intensified because of the situation and any mild culture shock may be expounded by the demands of learning and teaching in situ. It is therefore important to recognize the physical, mental, and emotional demands on students as created by our learning opportunities and assessments. As tutors we need to be sensitive and anticipate reactions and carefully intervene as necessary. Tutors can help structure the support students receive through regular dialogue, have impromptu informal discussions with groups and individuals, and use reflective fieldwork notebooks and journals to encourage personal reflection in order for students to make sense of their geographical encounters. This can lead to transformative learning as students share their thoughts, experiences, and emotions and co-create knowledge within their groups.

During the fieldtrip the daily reflective questions act as prompts for students to observe and record their thoughts in their reflective fieldwork diary. These questions provide a framework through which they can make sense of what they have experienced. With appropriate support and facilitation by tutors and by encouraging students to reflect on their experiences and emotions, students can develop strategies for managing the demands of the fieldtrip. Tutors can assist them by providing formative feedback and encouraging them to learn from their experiences and emotions to reinforce their learning.

Examples of daily reflective questions (summative):

- What are your first impressions of the city?
- What is meant by 'place' and what gives Barcelona its sense of place?
- How have your perceptions of Barcelona changed and why?

- Is the area of Las Ramblas primarily for tourists or for local people and does this cause any tension?

The self-reflective essay draws from comments made within the fieldwork notebook and is designed to allow students to reflect on the assessment activities that they participated in during the fieldtrip.

Self-reflective Essay (post-fieldwork, summative) Remit:–

Based on your experiences of preparing, leading, and participating in student-led tours and activities, reflect on your experiences of:

- Planning and delivering a presentation in the field
- Leading a field activity
- Participating in the field tours and activities of other students.

5.6 Student Responses to Assessment During Fieldwork

The assessment design complements the peer-teaching pedagogy and provides a structure to develop the student experience through the module. Assessment types are designed to facilitate the pedagogy, providing a framework for progressing the student academically, through formative and summative assessments, with feedback at each stage. Student responses to feedback and assessment are recorded in the fieldwork notebook, the reflective essay, and an end-of-module evaluation as well as through anecdotal comments.

The initial group pre-fieldtrip project proposal provides an initial focus. Group work can be challenging for some students as group dynamics can both encourage and also impede progress. Groups are formed around the interests of students around broad thematic topics:

> I think the way in which the groups were put together (by topic of interest) was a good way of avoiding self-selecting groups of friends. (Student #1, female, formal reflective essay)

Although not everyone had a positive experience:

> It became apparent that there was a poor dynamic within our group … I believe in the importance of research, preparation, and interaction, but found that other group members did not take the same approach. (Student #2, female, formal reflective essay)

Being in a classroom, the student groups are asked during each session to report back to the class verbally on their progress. This not only gives the tutor a chance to provide formative feedback but allows the students themselves to informally judge the progress made by other groups, compare their own progress, and learn from each other. Once submitted, the group pre-fieldtrip project assessment allows the tutor to deliver summative feedback and for groups then to act on the feedback in the form of 'feedforward' (Hirsch 2017) to further refine their project ideas.

During fieldwork, the role of the reconnaissance day is important for making the transition between the classroom and being in situ. For some students, the transition was smooth but for others there was a disparity between using online resources, books, and journals in comparison to the lived reality of being there. Even after bringing pre-prepared material to the field-site, for some students it became apparent that they had either over-prepared or under-prepared. This resulted in a reconfiguration of material in preparation for the assessed presentation and fieldwork activity, but this was not without consequence (see Fig. 5.1):

> The reconnaissance day was very valuable as it allowed us to map out our locations and create an itinerary with set times for each activity and part of the presentation. (Student #3, female, field notebook)

> After visiting the sites on the reconnaissance day we subsequently recognized the need to cut our scripts in half ... The need to cut down our scripts caused many problems as we each had different ideas about what should be cut down ... Upon reflection I can see that we were perhaps a little too ambitious …. (Student #4, female, field notebook)

Delivering the group field presentation and activity is the main focus. Some students relished the opportunity, whilst others expressed a degree of anxiety and trepidation. Very often students were making changes based on their new knowledge of their locations through being in situ:

> I was still doing last minute adjustments the night before just hoping to make sure everything went well, and not to let my group down, after all the weeks of study would come down to an hour and a half, make or break, pass or fail. (Student #5, male, field notebook)

Some students began to feel a weight of responsibility as they were accountable for the learning of their peers. The presentations deliver key information that students can include in their summative essay:

> I recognized the responsibility we held to the other students in providing a good presentation so that they would have useful grounds on which to base their own essays on the transformation of Barcelona. (Student #6, female, formal reflective essay)

Students are not only learning academic knowledge from their peers but they are also observing presentation styles and making critical judgements about their learning activities:

> I found that I became more analytical of the presentations before me. (Student #7, male, formal reflective essay)

> I did find it quite stressful with the timings, and compared myself to the previous presentation that seemed very neat and covered [the area] well, but didn't have such a long route as ours, so I felt pressured to get everyone round the tour in time to stop and let people take the surroundings in for themselves, I was also conscious of our tour feeling disjointed. (Student #6, female, formal reflective essay)

Being in the audience was not a passive experience. The students exercised a degree of mutual support and freely participated when asked to do so by the presenting group (see Fig. 5.2):

Fig. 5.1 Students preparing for their peer-led field presentation following a reconnaissance visit to their field sites, Barcelona

Fig. 5.2 Students presenting to their peers on the topic of gentrification and social exclusion. Protest banner in El Raval, Barcelona

> I initially anticipated the challenge in motivating each other to join in with group activities; however, we engaged with each other because we wanted each other to succeed. (Student #8, female, formal reflective essay)

The student audience plays an important part in providing formative feedback to the presenting students through questions and discussion. These thoughts are also recorded in their reflective diary and formal reflective essay. Although some groups performed better than others, what was surprising was that there emerged a competitive spirit between the groups:

> Some groups spoke very fast and sometimes very quiet which was frustrating as it was hard to write down points and it left gaps in my knowledge. However, I was able to fill these [gaps] during most activities where discussions were held as I was able to ask questions to complete my notes. (Student #9, male, formal reflective essay)

> Some activities in my opinion were a lot more educational than others with some having more emphasis on fun where I was engaged but didn't really seem to learn anything … we were concentrating more on winning. (Student #10, male, formal reflective essay)

Being in situ there is less control of the learning environment as students are immersed within the field and as such have to respond to challenges of dealing with the daily life of the city. Often groups had not considered this and were shocked when disruptive instances occurred:

> Presenting to a group of students [in situ] was difficult though. Constantly having to make sure everyone could hear: each environment presented different difficulties such as traffic noise, music, and other tourist groups. (Student #11, female, formal reflective essay)

> Local people and tourists did at one point become a problem for us, several walking through our group during one section of the presentation. I also noticed this happening to other groups … who had a couple sit down in the audience to listen to the presentation. (Student #12, female, formal reflective essay)

However, for some students the experience of delivering their presentations and being involved in an intensive learning experience led them to feel tired and under-prepared for assessment:

> I found that whilst working long hours on the two previous days and adding to our presentation in the evenings, that tiredness played a large part by the time that we had planned to deliver our presentation, unlike any other fieldtrip we had participated in. (Student #13, female, formal reflective essay)

The field notebook and self-reflective essay allows students to record their thoughts in situ and also to use the notebook as a source to construct the self-reflective essay. Both sources provide a unique insight into the student experience:

> [The notebook] worked very well, especially writing this brief reflection. I am able to read what I thought from the beginning of the fieldtrip to the end, looking at my thoughts, challenges we faces [sic] and how we overcame them. (Student #14, female, field notebook)

The use of reflective writing and getting into the habit of recording experiences was a challenge to some. The group were shown in the preparatory workshops how to write in a reflective style but for some it took a while to grasp:

> The reflective aspect of the module was difficult to engage with at first; however, after the first activity reflecting on my experiences and perspectives of various aspects of the city and the fieldtrip became much easier, allowing me to reflect on what I was learning. (Student #15, female, formal reflective essay)

The self-reflective essay encourages students to draw from academic discussions using pedagogic research papers. It not only contextualizes the student experience of delivering a group field presentation and activity but reveals experiences and assessed activities of other fieldtrips. The inclusion of pedagogic literature in the assessment allows them to make comparisons and reflect on their own performance and experiences:

> Hovorka and Wolf (2009) comment that when students are encouraged to take responsibility ... it increases self-confidence and the students are actively building up their own knowledge ... In my case I believe this was true, I was interested in the topic my own group had covered, but also interested to see what other topics had been covered and how the groups had constructed their own presentations and activities. (Student #16, male, formal reflective essay)

> [The reconnaissance visit] was eye-opening as we were shocked by the sheer size of the site compared to how we thought it was going to seem. Smith (2006) defines this perception as the 'fieldwork gaze' students have from researching an area before visiting. (Student #17, female, formal reflective essay)

The final assessment, an essay on the 'transformation of Barcelona', acts as a capstone to bring together aspects of geographical processes and further develops the academic content delivered during the student presentations. Students are aware of their role in the assessment process, not just researching for themselves but also instructing others:

> The way that we understand and experience the transfer of knowledge was changed by the fact that the people giving me knowledge were my peers. I was acutely aware of this whilst presenting my own work. (Student #6, female, formal reflective essay)

It is the integrated design of the assessments that enables students to receive these insights, complementing the peer-teaching pedagogy. The assessment is not replacing the pedagogy but enhancing it, structuring the student's development and acquisition of knowledge, skills, and experiences within the affective domain:

> When we went off in our individual activity groups for our group activity, all but one of the groups made the meeting point on time. There, we had to improvise for ten minutes or so and keep interacting with the class until the last group arrived. This is something I wasn't expecting and I was surprised at how we as a team reacted to this dilemma. We remained calm and kept the class discussing topics within the presentation. This provided the discovery of new skills in the area of public speaking including team working, taking initiative, creativity, leadership, and problem solving. (Student #18, female, formal reflective essay)

5.7 Discussion

International fieldwork places a variety of demands upon the students. It is important to be clear on what to assess, in this case the summative assessments are informed by formative assessment throughout the fieldtrip. Students can draw from their in situ experiences, personal reflections, and pedagogic literature to gain a deeper understanding and a sense of place. As such, assessment is integral to fieldwork.

The use of formative assessment plays a key part in the success of the fieldwork. Through the preparatory workshops students have regular informal conversations with tutors and their peers. It should not be understated the influence that peer–peer learning has on the co-production of knowledge and understanding. Through providing verbal reports in class sharing progress and ideas, they gain confidence in talking in front of their peers and engaging in discussion with others. They are able to monitor their own learning and progress through regular engagement as a community of learners. Fieldwork is a process and not a product in so far as students are constantly experimenting and discovering new ideas which impact on summative activities. Fieldwork is a voyage of discovery and students should be allowed to experiment with ideas and discover potential issues for themselves.

The approach to assessment taken during the module is based on tutors assessing students' academic performance through an understanding of the geography of Barcelona, and also on the positionality and experiences of fieldwork through a sense of place and emotional geographies. The two are inseparable as one reinforces the other. This reaffirms the idea that assessment and feedback are inextricably linked. There is an ongoing dialogue throughout the module which is integral to the teaching and learning strategy. Assessment, by various means, allows the facilitation of feedback which then informs the next stage of the assessment. For example, workshops inform the project proposal, feedback then informs the design of field presentation, and peer–peer teaching informs written essays. It can therefore be stated that 'assessment' is occurring throughout the module as part of tutor involvement, including informal discussions as well as formal feedback. The concept of students as learning partners or co-creators of knowledge are purposefully designed into the assessment tasks and learning situations.

Allowing students to demonstrate greater responsibility for their learning and that of others needs careful management. Tutors need to reaffirm boundaries as to what is achievable and what is expected of the student groups. As such, feedback can be regarded by students as positive as well as negative, but both need to be constructive in order to facilitate learning and to feedforward (Hirsch 2017). Feedback is therefore considered as an active dialogue between tutor and student and between students. As tutors, what we say and do can influence situations and outcomes (Cotton et al. 2013). Each year tutors give a health and safety talk before the fieldtrip including aspects relating to logistics, what to bring, behavioral expectations, and personal safety. It is also discussed where to go and where not to go. Students are asked to show caution when visiting El Raval, particularly at night (Degen 2017). However, it became apparent that some students became overly concerned and sought reassurance from tutors. It is therefore necessary to be mindful of personal circumstances (Glass 2014) and show empathy, maintaining reassurance and a positive attitude which can increase levels of openness, trust, and mutual respect. This is also reinforced by creating responsive and compassionate learning and teaching strategies and assessment, therefore creating a compassionate pedagogy, particularly through critical self-reflection (Hao 2011).

There is no evidence in the field diaries or reflective essays that students felt disadvantaged by student-led teaching. Some did not particularly enjoy the experience, through not being interested in the subject context or not being able to hear clearly due to background noise. Although students appreciated the professionalism of staff-led instruction, most preferred being student-led. Individual interest arose out of finding out what other students had been doing and, in particular, in the creativity of the other field activities. There is also a sense of informality to learning from students as staff-led is perceived as being more formal, structured, and regimented. Student-led teaching was viewed as being more accessible, using language that they could relate to, but this may indicate that they are not being academically challenged or are receiving less conceptual knowledge than a tutor would deliver. Yet the informal setting suggests that students enjoyed the experience more than being in a classroom (see Fig. 5.3).

Fig. 5.3 Students studying the conflicts between local inhabitants and tourists at the popular Parc Guell attraction, Vallcarca district, Barcelona

It is clear that different assessment items are created to emphasize different learning outcomes, but they also capture distinct aspects of the fieldtrip experience. Field diaries typically contain greater levels of emotion whereas reflective essays are more 'sanitized' and use language and quotes that are written for the intended audience, the tutors (Marvell and Simm 2018). In contrast, the field diaries contain a greater range of phrases that reveal achievement, frustration, fear, fatigue, and other raw emotions that do not necessarily appear in the reflective essays. The diary is used to reflect upon the actual experiences and skills used and tends to be open and honest but with less depth of critique.

The reflective essays tend to adopt a more formal academic writing style. They are more considered and refined with less sense of spontaneity. Students often make links to the pedagogic literature, imitating the formal style of written argument and citing relevant and supporting statements. They also include phrases and quotes from their diaries to illustrate their experiences. The more able students attempt to make sense of what has been experienced. Students typically have no background in educational theory and instead rely on the pedagogy in the research papers that are suggested by tutors. The reflective essays seek to provide a greater depth and analysis of the students' own performances, although a minority of students seek to blame other people or situations rather than focus on their own. The academic writing style in the reflective essays is a result of the task and nature of the assessment: the expectation to include pedagogic literature and theory, student response to the marking criteria, and the perceived expectations of tutors. We have to remember that the assessment tasks encourage self-reflection of the learning process rather than explicitly considering the geographical nature of place and environment as a vehicle or backdrop for this process. Therefore, fieldwork diaries and reflective essays need to be treated and assessed differently.

5.8 Future Recommendations

From this experience there are many ways that student-led fieldwork could evolve. Greater integration of mobile digital technology can allow students to have greater access to a range of digital resources whilst in situ. Rather than recording ideas in a paper diary, the format could be electronic and appear as a personal blog. The nature of the comments may be such that the blog remains personal rather than public. Groups could be tasked to produce videos or vlogs that capture their engagement with the study areas. The challenges of fieldwork in a post-COVID world may be facilitated by greater use of technology. Group presentations could be recorded and used as resources, or the content could be streamed in real time to allow others in different locations to interact with the fieldwork team. France and Ribchester (2004) utilize web technology as a platform to assess students writing up a field-based research project as a website. This can potentially add additional functionality compared with a traditional written essay or report as a range of media can be embedded. Another application of internet technology as a form of learning and assessment is through the use of blogs, encouraging a community of learning and dialogues between learners as well as tutors (Marvell 2018). Assessing written blogs can ensure a higher level of student engagement as students predominantly focus on activities that are linked to assessment (Morris et al. 2019). As students are often assessment focused, rather than working against this, assessments should be clear and creative in order to enhance learning (Worth 2014). Creative methods can be applied to capture the experiences of being on fieldwork. For example, Mavroudi and Jöns (2011) apply the use of video documentaries in human geography field courses that can be used to record personal reflections as well as digital presentations of research.

The role of students as 'scholars' through peer-teaching during fieldwork (Marvell et al. 2013) can be further enhanced by including peer assessment. This would extend the role and responsibility of the students to not only peer-teach but also be responsible for rewarding academic success. It would allow students to provide formative feedback to their peers (El-Mowafy 2014). However, this would need to be carefully integrated as summative assessment so that students did not just reward creative and entertaining presentations with higher marks but also recognize academic scholarship, knowledge, and criticality.

Although the ideas presented in this chapter have been used with students in higher education, student-led teaching can be adapted for use in other phases of education. Clark and Spamer (2017) adapted our fieldwork pedagogy for a group of 16–18 year old pre-university students. The approach to fieldwork enabled students to become progressively reflective and ask deeper questions about their surroundings. One student stated the approach made them look and ask questions, 'I find myself walking through town and wondering how it came to be like this' (Clark and Spamer 2017, p. 62). Giving students opportunities to explore ideas as a group and present to each other is not a new concept but as tutors we are often reluctant to transfer power and responsibility to students.

5.9 Conclusion

The role of assessment and feedback are closely aligned as they facilitate the students' learning journey. Fieldwork often presents a series of challenges and opportunities to explore a range of pedagogies and possibilities. This chapter demonstrates that assessment can play an important role in shaping the teaching and learning strategy that underpins the student experience of being on fieldwork. It is true to say that in the design of the assessment the concept of the activity came first and the assessment framework then followed, but they are closely aligned. Through constructive alignment it is possible to devise an assessment framework that is incremental in nature, offering a variety of feedback opportunities throughout the fieldtrip. Encouraging students to become reflective learners is an essential step in achieving this; however, they need to be taught how to be reflective and then how to apply their thoughts to the assessment tasks. By allowing students to explore ideas and take on responsibilities for the teaching of others develops a range of skills and experiences that cannot be easily replicated within the classroom. It is only when students reflect on their experiences and assessments that they begin to appreciate what they have achieved:

> At the time of the presentation I did not appreciate how many skills I was developing and using. It was only when you reflect on your experience that it becomes obvious how difficult the task actually was. (Student #19, female, formal reflective essay)

References

Arrowsmith C, Bagoly-Simó P, Finchum A, Oda K, Pawson E (2011) Student employability and its implications for geography curricula and learning practices. J Geogr High Educ 35(3):365–377. https://doi.org/10.1080/03098265.2011.563379

Bennett RE (2011) Formative assessment: a critical review. Assess Educ Princ Pol Pract 18(1):5–25. https://doi.org/10.1080/0969594X.2010.513678

Biggs J (2003) Aligning teaching for constructive learning. The Higher Education Academy, York. https://www.advance-he.ac.uk/knowledge-hub/aligning-teaching-constructing-learning

Black P, Wiliam D (2012) Assessment for learning in the classroom. In: Gardner J (ed) Assessment and learning. Sage, London, pp 19–35

Bloom BS (1969) Some theoretical issues relating to educational evaluation. In: Tyler RW (ed) Educational evaluation: new roles, new means: the 63rd yearbook of the National Society for the Study of Education (Part II). University of Chicago Press, Chicago, pp 26–50

Clark C, Spamer K (2017) Unravelling the palimpsest of Hessle Road, Hull. Teach Geogr 42(2):60–62

Coe NM, Smyth FM (2010) Students as tour guides: innovation in fieldwork assessment. J Geogr High Educ 34(1):125–139. https://doi.org/10.1080/03098260902954095

Cotton D, Winter J, Bailey I (2013) Researching the hidden curriculum: intentional and unintended messages. J Geogr High Educ 37(2):192–203. https://doi.org/10.1080/03098265.2012.733684

Cuddapah JL, Clayton CD (2011) Using Wenger's communities of practice to explore a new teacher cohort. J Teach Educ 62(1):62–75. https://doi.org/10.1177/0022487110377507

Dann R (2014) Assessment as learning: blurring the boundaries of assessment and learning for theory, policy and practice. Assess Educ Princ Pol Pract 21(2):149–166. https://doi.org/10.1080/0969594X.2014.898128

Degen M (2017) Urban regeneration and "resistance of place": foregrounding time and experience. Space Cult 20(2):141–155. https://doi.org/10.1177/1206331217697136

Dummer TJB, Cook IG, Parker SL, Barrett GA, Hull AP (2008) Promoting and assessing 'deep learning' in geography fieldwork: an evaluation of reflective field diaries. J Geogr High Educ 32(3):459–479. https://doi.org/10.1080/03098260701728484

El-Mowafy A (2014) Using peer assessment of fieldwork to enhance students' practical training. Assess Eval High Educ 39(2):223–241. https://doi.org/10.1080/02602938.2013.820823

France D, Ribchester C (2004) Producing websites for assessment: a case study from a level 1 fieldwork module. J Geogr High Educ 28(1):49–62. https://doi.org/10.1080/0309826042000198639

Francis RA, Millington JDA, Cederlöf G (2019) Undergraduate student perceptions of assessment and feedback practice: fostering agency and dialogue. J Geogr High Educ 43(4):468–485. https://doi.org/10.1080/03098265.2019.1660867

Fuller I, Edmondson S, France D, Higgitt D, Ratinen I (2006) International perspectives on the effectiveness of geography fieldwork for learning. J Geogr High Educ 30(1):89–101. https://doi.org/10.1080/03098260500499667

Glass M (2014) Encouraging reflexivity in urban geography fieldwork: study abroad experiences in Singapore and Malaysia. J Geogr High Educ 38(1):69–85. https://doi.org/10.1080/03098265.2013.836625

Graven M, Lerman S (2003) Book review: Wenger, E. (1998) Communities of practice: learning, meaning and identity. J Math Teach Educ 6(2):185–194. https://doi.org/10.1023/A:1023947624004

Hao RN (2011) Critical compassionate pedagogy and the teacher's role in first-generation student success. New Dir Teach Learn 2011(127):91–98. https://doi.org/10.1002/tl.460

Hargreaves E (2005) Assessment for learning? Thinking outside the (black) box. Camb J Educ 35(2):213–224. https://doi.org/10.1080/03057640500146880

Higgitt D (2014) Disruptive moments. J Geogr High Educ 38(1):1–6. https://doi.org/10.1080/03098265.2014.885172

Hill J, Healey RL, West H, Déry C (2021) Pedagogic partnership in higher education: encountering emotion in learning and enhancing student wellbeing. J Geogr High Educ 45(2):167–185. https://doi.org/10.1080/03098265.2019.1661366

Hill J, Thomas G, Diaz A, Simm D (2016) Borderland spaces for learning partnership: opportunities, benefits and challenges. J Geogr High Educ 40(3):375–393. https://doi.org/10.1080/03098265.2016.1144728

Hill J, Walkington H, King H (2018) Geographers and the scholarship of teaching and learning. J Geogr High Educ 42(4):557–572. https://doi.org/10.1080/03098265.2018.1515188

Hindle P (2000) Developing employment interview and interviewing skills in small-group project work. J Geogr High Educ 24(1):29–36. https://doi.org/10.1080/03098260085117

Hirsch J (2017) The feedback fix: dump the past, embrace the future and lead the way to change. Rowman & Littlefield, Lanham

Holton M (2017) "It was amazing to see our projects come to life!" Developing affective learning during geography fieldwork through tropophilia. J Geogr High Educ 41(2):198–212. https://doi.org/10.1080/03098265.2017.1290592

Hovorka AJ, Wolf PA (2009) Activating the classroom: geographical fieldwork as pedagogical practice. J Geogr High Educ 33(1):89–102. https://doi.org/10.1080/03098260802276383

Ito H, Igano C (2021) International fieldwork as skills development: an exploratory study. J Geogr High Educ 45(3):417–434. https://doi.org/10.1080/03098265.2020.1836481

Kent M, Gilbertson DD, Hunt CO (1997) Fieldwork in geography teaching: a critical review of the literature and approaches. J Geogr in High Educ 21(3):313–332. https://doi.org/10.1080/03098269708725439

Kim M (2020) Developing pre-service teachers' fieldwork pedagogical and content knowledge through designing enquiry-based fieldwork. J Geogr High Educ. https://doi.org/10.1080/03098265.2020.1849065

Kolb DA (1984) Experiential learning: experience as the source of learning and development. Prentice-Hall, Englewood Cliffs

Krathwohl DR (2002) A revision of Bloom's taxonomy: an overview. Theor Pract 41(4):212–218. https://doi.org/10.1207/s15430421tip4104_2

Krathwohl DR, Bloom BS, Masia BB (1964) Taxonomy of educational objectives, handbook II: affective domain. David McKay, New York

Magolda MBB (2014) Self-authorship. New Dir High Educ 2014(166):25–33. https://doi.org/10.1002/he.20092

Marvell A (2008) Student-led presentations in situ: the challenges to presenting on the edge of a volcano. J Geogr High Educ 32(2):321–335. https://doi.org/10.1080/03098260701514173

Marvell A (2018) Student experiences of facilitating knowledge exchange: developing an understanding of responsible events through blog writing. J Hospit Leisure Sport Tourism Educ 23:1–9. https://doi.org/10.1016/j.jhlste.2018.04.002

Marvell A, Simm D (2018) Emotional geographies experienced during international fieldwork: an evaluation of teaching and learning strategies for reflective assessment. J Geogr High Educ 42(4):515–530. https://doi.org/10.1080/03098265.2018.1460806

Marvell A, Simm D, Schaaf R, Harper R (2013) Students as scholars: evaluating student-led learning and teaching during fieldwork. J Geogr High Educ 37(4):547–566. https://doi.org/10.1080/03098265.2013.811638
Mavroudi E, Jöns H (2011) Video documentaries in the assessment of human geography field courses. J Geogr High Educ 35(4):579–598. https://doi.org/10.1080/03098265.2011.559578
McCormack DP (2008) Geographies for moving bodies: thinking, dancing, spaces. Geogr Compass 2(6):1822–1836. https://doi.org/10.1111/j.1749-8198.2008.00159.x
McGuinness M, Simm D (2005) Going global? Long-haul fieldwork in undergraduate geography. J Geogr High Educ 29(2):241–253. https://doi.org/10.1080/03098260500130478
Mezirow J (2000) Learning as transformation: critical perspectives on a theory in progress. Jossey Bass, San Francisco
Moon J (1999) Reflection in learning and professional development. Kogan Paul, London
Morris NJ, Christie H, Barber J (2019) 'It's one of the first times I've felt fully engaged': developing student engagement using blogging as a form of assessment. J Geogr High Educ 43(3):343–361. https://doi.org/10.1080/03098265.2019.1612862
Mortiboys A (2012) Teaching with emotional intelligence, 2nd edn. Routledge, London
Nicholson DT (2011) Embedding research in a field-based module through peer review and assessment for learning. J Geogr High Educ 35(4):529–549. https://doi.org/10.1080/03098265.2011.552104
Pawson E, Fournier E, Haigh M, Muniz O, Trafford J, Vajoczki S (2006) Problem-based learning in geography: towards a critical assessment of its purposes, benefits and risks. J Geogr High Educ 30(1):103–116. https://doi.org/10.1080/03098260500499709
Peasland EL, Henri DC, Morrell LJ, Scott GW (2019) The influence of fieldwork design on student perceptions of skills development during field courses. Int J Sci Educ 41(17):2369–2388. https://doi.org/10.1080/09500693.2019.1679906
Punch S (2012) Hidden struggles of fieldwork: exploring the role and use of field diaries. Emot Space Soc 5(2):86–93. https://doi.org/10.1016/j.emospa.2010.09.005
Scheyvens R, Wild K, Overton J (2003) International students pursuing postgraduate study in geography: impediments to their learning experiences. J Geogr High Educ 27(3):309–323. https://doi.org/10.1080/0309826032000145070
Scriven M (1967) The methodology of evaluation. In: Tyler RW, Gagné RM, Scriven M (eds) Perspectives of curriculum evaluation. Rand McNally, Chicago, pp 39–83
Simm D, Marvell A (2015) Gaining a "sense of place": students' affective experiences of place leading to transformative learning on international fieldwork. J Geogr High Educ 39(4):595–616. https://doi.org/10.1080/03098265.2015.1084608
Smith FM (2006) Euro-commentary: encountering Europe through fieldwork. Eur Urban Reg Stud 13(1):77–82. https://doi.org/10.1177/0969776406058957
Taras M (2005) Assessment—summative and formative—some theoretical reflections. Br J Educ Stud 53(4):466–478. https://doi.org/10.1111/j.1467-8527.2005.00307.x
Thrift N (2007) Non-representational theory: space, politics, affect. Routledge, Abingdon
Tinto V (2000) Learning better together: the impact of learning communities on student access in higher education. J Institutional Res 9(1):48–53
Trigwell K, Prosser M (2014) Qualitative variation in constructive alignment in curriculum design. High Educ 67:141–154. https://doi.org/10.1007/s10734-013-9701-1
Vygotsky LS (1978) Interaction between learning and development. In: Cole M, John-Steiner V, Scribner S, Souberman E (eds) Mind in society: the development of higher psychological processes. Harvard University Press, Cambridge, pp 79–91
Wenger E (2001) Communities of practice: learning, meaning, and identity. Cambridge University Press, Cambridge
Widdowfield R (2000) The place of emotions in academic research. Area 32(2):199–208
Worth N (2014) Student-focused assessment criteria: thinking through best practice. J Geogr High Educ 38(3):361–372. https://doi.org/10.1080/03098265.2014.919441

Alan Marvell is a Senior Lecturer in Events Management and Course Leader for the Doctorate in Business Administration at the University of Gloucestershire, UK. His Ph.D. is in Geography, and he has an interest in pedagogy and experiential learning. He has served as a Director and Trustee of the Geographical Association, UK and is an active member of the Geography and Education Research Group of the Royal Geographical Society. Alan is on the Editorial Board of the *Journal of Geography in Higher Education* and the *Journal of Hospitality, Leisure, Sport and Tourism Education*. He is also a Senior Fellow of the Higher Education Academy (Advance HE, UK) and was awarded a National Teaching Fellowship in 2021.

David Simm is a Senior Lecturer in Physical Geography at Bath Spa University, UK. With extensive experience in various HE roles including Learning and Teaching Co-ordinator, Subject Leader (Geography), and Course Leader (Development Geography), he has interests in problem-based and experiential learning, particularly though international fieldwork, devising innovative fieldwork courses involving student peer-led teaching and learning. He has served on the committee of the former Higher Education Research Group (now the Geography and Education Research Group) of the Royal Geographical Society, and serves on the Editorial Board of the *Journal of Geography in Higher Education*. He is a Senior Fellow of the Higher Education Academy (Advance HE, UK) and he was awarded a National Teaching Fellowship in 2020.

Chapter 6
Inquiry-Based Fieldwork Assessment *for* and *as* Learning in Geography

Chew-Hung Chang and Phoebe Ow

Abstract In balancing the role of assessment as an integral part of the curriculum and teaching process and that of a measurement and reporting tool, practitioners are often challenged to design good assessment tasks that fulfil these purposes as well as developing cognitive skills and abilities. While there are many research studies on inquiry in fieldwork and on assessment in geography education, there is little recognition on how the whole process of inquiry acts as a form of fieldwork assessment within geographical education. This chapter proposes that inquiry can be a mode of assessment *for* and *as* fieldwork and need not be solely administered at the end of fieldwork activities.

Keywords Inquiry · Fieldwork · Summative assessment · Formative assessment

6.1 Introduction

Geography is a crucial school subject for twenty-first century children in an interconnected world. A good hallmark of geographical education includes fieldwork which requires students to apply their knowledge in investigating a geographical phenomenon (Phillips and Johns 2012). This process allows students to develop a holistic and synergistic understanding of geographical issues (Kent et al. 1997) and provide 'a means towards examining the relationships between people and their environments' (Katz 2009, p. 251). Students learn what it means to live sustainably in this world in relation to other people, where human relationships and their responsibilities to both the natural environment and others are better understood. Indeed, it

C.-H. Chang (✉) · P. Ow
Nanyang Technological University, National Institute of Education, Singapore, Singapore
e-mail: chewhung.chang@nie.edu.sg

P. Ow
e-mail: phoebe.ow@nie.edu.sg

T. Bourke et al. (eds.), *Assessment in Geographical Education: An International Perspective*, Key Challenges in Geography,
https://doi.org/10.1007/978-3-030-95139-9_6

takes students' learning beyond knowing, to doing and to being as well as learning to live together; it helps people learn how to exist harmoniously with all living species (International Geographical Union—Commission on Geographical Education 2016, p. 5).

While the intention to help students develop these learning outcomes is clear, the way we assess these students in attaining the above outcomes is something we must not overlook. Arguments in assessment have called for a paradigmatic shift, with a focus on supporting students' learning instead of simply sorting and selecting students (Delandshere 2002). Similarly, for assessments in fieldwork, emphasis has shifted towards active engagement and connections in the learning experienced by students. Posited by Wells (1999, p. 121), inquiry is seen 'as the construction of new understandings and a stance towards experiences and ideas – a willingness to wonder, ask questions and seek to understand by collaborating with others'. In this sense, adopting an inquiry-based approach to assessment in an authentic setting not only enables students to develop deeper geographical understanding, but offers a modality of assessment for us to know if students have learnt what was intended along the way.

Within fieldwork, the inquiry-based approach is a process based on hands-on, real life activities to provide students with the opportunity to engage in uncovering unknown causal relations by constructing and investigating hypotheses through conducting experiments and observations (Pedaste et al. 2012). Assessments in fieldwork must be able to assess these qualities by demonstrating properly what and how students have learned in the field. Additionally, students should go deeper beyond learning and knowing towards dispositional changes in the way they see themselves as part of the human–environment relationship. Abilities in showcasing connections and reflexivity needs to be further assessed if we want to help our students attain higher levels and broader learning outcomes. Assessment must therefore go forward in considering the overall qualities of a student instead of plainly distinguishing their strengths and weaknesses in particular knowledge or skills.

Traditionally, assessment is thought of as a summative construct to determine if students have learned what we intend for them to learn. However, an assessment of what and how much each student has learned is not only necessarily carried out in the end; it can also be integrated as part of the learning activity itself (Chang and Seow 2018). Commonly seen as a practice separate from the learning and teaching process, we argue that assessment in and of itself is part of the learning process. In this chapter, we contend that inquiry as a learning process is a mode for assessment in fieldwork.

In this chapter, we provide an overview of the assessment literature pertaining to fieldwork inquiry. Although there has been a shift for assessment practices to center around identity building and support for learning, fieldwork assessments in geography education continue to lean towards traditional modes such as summative and standardized assessment practices. Thereafter, we discuss reasons as to why traditional modes of assessments are still mainly practiced despite their limitations within fieldwork. To push the argument further towards why and how the inquiry process is itself assessment for and of good fieldwork, epistemological underpinnings of the inquiry process are provided. Finally, we discuss how inquiry is implemented as a form of assessment for fieldwork by presenting an example in the Singaporean context.

6.2 Summative Versus Formative Approaches in Fieldwork Assessment

Summative assessments are the most common assessment approach used in geographical inquiry fieldwork. They include standardized assignments, such as tests at the end of fieldwork or final projects often accompanied by student presentations of what they have found. These types of assessments are frequently used together in fieldwork activities. However, since the 1900s, education research and pedagogy has seen a shift in modes of assessment (Brown 1969; Moore 2003; Looney 2011). This shift is seen in the changing emphasis from summative to formative assessment (Kent et al. 1997; Delandshere 2002; Moore 2003). The change can be largely attributed towards keeping up with current global trends and issues that not only require content knowledge, but equally require skills to comprehend a planet that is ever-changing. In the context of geography education, goals have evolved from merely the study of the natural environment to developing the political, social, and cultural understanding of present and future activities and environments (International Geographical Union—Commision on Geographical Education 2016). The move from passive to active learning reflects the recognition of deeper engagement for students to not only learn, but also to build relationships and complex understandings of the environment around them.

Similarly, within geographical fieldwork, active learning is being foregrounded. Research indicates the shift away from traditional 'cook tours' and memory-dependent fieldwork assessments to fieldwork and assessments that critically engage students to create opportunities for learning holistically (Preston 2016; Kho and Parker 2010; Das and Chatterjea 2018). As such, there has been a rise in inquiry-based fieldwork learning (Lim 2018; Casinader and Kidman 2018) where deep learning and affective engagement for students should be cultivated as well as enabling the achievement of learning goals and outcomes for students. Formative assessments that rely on information gathered during the assessment (learning) process is thus emerging as one of the principal forms of assessment for students in fieldwork.

According to Looney (2011), if assessment provides active and interactive learning processes, it is counted as formative. Through formative assessments, students are actively engaged in the process of learning and inquiry. A well-designed formative assessment in fieldwork enables students to not only dwell on ideas emerging in the academic literature but also in the field. Reflexivity is encouraged and promoted in all modes of assessment, but it is especially crucial for practices within the field so students can connect areas of knowledge, creating pathways for analysis and evaluation. This is because textbook scenarios simply do not exist in the field. Students are not operating within known parameters but, rather, in real-world contexts where elements of the environment can change over short periods of time. Fieldwork inquiry assessment must therefore be able to generate reflexivity so that students can figure out what to do, generating alternative hypotheses when situations change. Through such means, students are encouraged to go beyond observations and content knowledge, enabling cognitive development to occur. Without this, missed opportunities in linking everyday knowledge to powerful knowledge—the ability to extend beyond current knowledge and doing—arise (Butt 2017).

Additionally, feedback channels take a step further in providing reflexivity. Promoted by Schön (1987), it is insufficient to just reflect; feedback helps to create a double loop in reflexivity that takes the student into a deeper process of learning and cognitive skills. Besides creating linkages and connections to knowledge of real-world situations, students must also be able to recontextualize pre-set ideas to further knowledge. As elaborated by Brockbank and McGill (1998), the goal of any assessment approach is to inculcate deep and reflective learning by enabling learners to contemplate issues whilst obtaining feedback and insight that enables them to reflect longer to comprehend more deeply. Hence, to provide deeper learning and reflexivity in fieldwork, feedback channels must be properly incorporated within fieldwork assessments to promote higher order thinking. This can be achieved by setting proper guidelines, objectives, and design for proper formative feedback for both students and teachers.

One formative assessment strategy that has been promoted as a means of reflexivity and providing feedback channels in geography fieldwork is reflective diaries/journals. For example, in Moon's (2005) study, reflective diaries were used to foster activities for reconstructing knowledge and ideas. Students learned to generate new knowledge and reconstructed potential misconceptions held. In another study, Dummer et al. (2008) highlighted how students found their journal task to be refreshing, new, and challenging; they enjoyed the opportunity to personalize and individualize their improved academic writing. A more recent study by Simms and Shanahan (2019) reported on the implementation of individual journals to capture students' personal reflections during participation in a clean-up program. Results showed how students demonstrated 'transferrable' knowledge in that they were able to apply information to new situations (Simms and Shanahan 2019). Furthermore, students demonstrated criticality by examining multiple perspectives (Simms and Shanahan 2019). Based on the examples described, reflective diaries/journals, therefore, not only resolve students' misconceptions but arouse students' interest in geographical issues and developing higher order thinking skills such as criticality. These studies show that reflective journals can be used by both students and teachers to assess learning and capture the reflexivity and development of student thinking through different stages of fieldwork inquiry.

Despite the emphasis on formative assessments within fieldwork, traditional modes of assessment remain predominant. Literature on fieldwork activities often reflect ungraded formative assessments whilst summative assessments are ranked and considered for the overall module grade. This leads students to think that formative assessments are unimportant (Kent et al. 1997), resulting in a lack of priority and sloppy completion of formative assessment that affects assessment reliability. Furthermore, there is a pattern of teachers reverting to traditional modes of standardized assessments (Kent et al. 1997; Moore 2003). The next section will provide reasonings and limitations as to why traditional forms of assessments and tests are still preferred.

6.3 Problems in Assessments *of* Fieldwork

For effective assessment to take place, it should be integrated across the whole teaching and learning cycle from the early stages of planning (Rogers 2011). However, realizing this aim is often a challenge for time and resource poor teachers (Das and Chatterjea 2018; Summers et al. 2005; Hill and Dyment 2016). For example, Kho and Parker (2010) reported how Singaporean teachers were burdened with administrative tasks associated with planning fieldwork activities, leaving little time to plan and execute effective learning and assessment. In terms of resources, Baidoo-Anu et al. (2019) indicated how restrictive budgets and lack of personnel adversely affect opportunities to conduct fieldwork with students.

Some teachers are also less confident in their content knowledge for fieldwork inquiry assessment (Preston 2016), especially where formative assessment should be incorporated. According to Looney (2011), there is a certain degree of subjectivity attached to formative assessment with many teachers constructing assessments based on their own ideas and assumptions. Collins (1981, cited in Meyer et al. 2011) coins this phenomenon as interpretive flexibility which results in a 'less than emphatic agreement to meaning' of the assessment (Casinader and Kidman 2018, p. 11; Ferretti 2018; Rawling 2007). Moreover, there tends to be a lack of guidelines and knowledge on how to structure formative assessments. This makes it difficult for the teacher to gauge the learning of the student (Dummer et al. 2008; Brannon 2014). For these reasons, teachers often revert to traditional modes of assessment and pedagogies.

Although summative assessments are useful for measuring student outcomes, standardized assessments are insufficient in assessing skill development and are limited in assessing students' behaviors and attitudes (Looney 2011). As we have argued that the goal of a good geography education is for students to become custodians of our common environmental future, the outcomes of traditional summative assessment that tend to be related to specific skills and knowledge may not be aligned to this goal. Indeed, a one-size-fits-all standardized test, for example, will be unable to consider the differences between students in terms of individual dispositions and beliefs. We are also unable to ascertain if students can function as part of a team or are able to apply their skills and knowledge to solve authentic real-world problems. There is also the matter of the inability to determine the students' affective learning outcomes.

Many of the issues raised above—lack of resources, time, and content knowledge—arise due to the fact that assessment and curricula are seen as essentially separate instead of being a focal dimension to all processes in geographical fieldwork (Lambert and Reiss 2014). We assert that if we use inquiry, both as an instructional approach and mode of assessment, we will be able to promote good geographical learning in fieldwork and greatly reduce the burdens from the problems mentioned above.

6.4 Assessment *for* and *as* Fieldwork Inquiry

Geography focuses on knowing many theories and facts about intrinsic processes and patterns that allow individuals to comprehend a planet that is ever-changing (International Geographical Union—Commision on Geographical Education 2016). Inquiry as instruction and assessment enables the teacher to incorporate product and process to evaluate student outcomes according to curriculum goals. In doing so, it allows teachers to design meaningful fieldwork for students to gain deeper learning and skills that eventually influence their decisions, actions, and behaviors. To further break down the process of assessment in inquiry, scholars have developed approaches that highlight different stages of inquiry. One framework illustrates inquiry in phases: orientation, conceptualization, investigation, conclusion, and discussion (Pedaste et al. 2015). According to Pedaste et al. (2015), these phases are not linear; their pathways intersecting depending on how students dissect the research topic at hand. In every phase of inquiry, students acquire and display different skills which can be assessed. Not only do these phases act as a step-by-step guide for teachers to plan fieldwork curricula, but they also provide opportunities for teachers to understand and evaluate students' thinking and learning during the fieldwork process. Indeed, from an inquiry perspective, what is to be learned is as important as how it is being learned. In this sense, assessment *for* and *as* fieldwork is an ongoing inquiry process to accomplish student outcomes. The case example below will illustrate the use of a 'Geographical Investigation' schema in Singapore for inquiry-based fieldwork.

6.5 A Case Example

In this section, we use an example of Geographical Investigations in Singapore to elucidate some of the features described above. Kho and Parker (2010) have indicated how in Singapore, executing fieldwork is rooted in the processes of inquiry accompanied by well-established planning. This section will first highlight how inquiry-based fieldwork assessments are conducted as part of Geographical Investigations (GI) in the geography Singaporean curriculum (Curriculum Planning & Development Division 2021) before outlining a case example about how pre-service teachers are prepared to use this approach.

6.5.1 Geographical Investigations

In Singapore's geography curriculum, Geographical Investigations (GI)—a modified version of Roberts's (2013) inquiry framework—is adopted to engage students in criticality, reflective thinking, independent learning (Curriculum Planning & Development Division 2014), and inquiry for geographical learning. GI is a 'guided inquiry' process designed for lower secondary learners who have not had exposure to geography as a subject (Curriculum Planning & Development Division 2014). It comprises four stages which include formulating the GI question, data collection, exercising reasoning, and reflecting on the investigation process where student learning and skills are assessed at each stage. Therefore, assessment *for* and *as* inquiry that includes opportunities for reflexivity and feedback can be easily incorporated into the planning and design of fieldwork activities.

In GI, students are required to investigate a geographical issue related to specific topics in the geography syllabus. Each geographical issue is framed around a GI question that students first formulate, followed by collecting, selecting, and presenting relevant data and information from their field investigation to answer the inquiry question they generated. An example of this inquiry question is depicted under the topic of water, where students are required to frame their inquiry around the core questions of:

- What is water and where is water found?
- Why is water available on Earth?
- What relationships does water have with (i) the environment and (ii) people?
- How can these relationships be sustainably managed? (Curriculum Planning & Development Division 2021, pp. 13–16)

Consequently, students are required to interpret the data they have collected to offer possible solutions to the GI question (Lim 2018). This provides the schema against which the students will execute the inquiry fieldwork, as well as the schema for teachers to assess what and how students have learned at each stage.

6.5.2 Case Example: Sustainability Learning Lab, an Example of Embedding Formative Learning Through Inquiry in Fieldwork in Singapore

To prepare and conduct GI assessments related to the topic of Sustainability, a core component of the Singaporean geography curriculum, a Sustainability Learning Lab (SLL) was set up to provide pre-service teachers with the necessary tools and experience (Chang et al. 2018) for conducting fieldwork *for* and *as* assessment. The virtual laboratory provides user-friendly, informative websites with access to pertinent data and multimedia information sources. Examples include lesson plans, fieldwork ideas, and fun and interactive classroom activities. The SLL also supports real, outdoor learning labs, such as the National Institute of Education (NIE) raingarden and the Jurong Eco Garden, that are instrumented to provide real-time and historical environmental data (e.g., meteorology, air quality, water, etc.). Equipped with these resources, this allows the SSL to prepare pre-service teachers using the GI approach to formulate suitable inquiry processes and assessments for themselves and their future students to experience learning outcomes that go beyond the knowledge domain. The following is an example of this approach using the subheadings of doing the fieldwork, writing/reflecting about the fieldwork, and, lastly, relating the formative processes to the final product, the summative assessment presentation.

6.5.2.1 Doing the Fieldwork—Inquiry *as* Learning

At the SLL, pre-service teachers participate in a GI fieldwork simulation. As a small team, pre-service teachers collect and analyze biological/meteorological data from several locations at NIE. Such data include water quality parameters such as temperature, dissolved oxygen, conductivity, pH, turbidity, and fluorescence (chlorophyll *a*). Data are obtained in numerical and visual forms using geospatial technologies and recording devices (Chang 2018). During the data collection processes, the lived experiences of the pre-service teachers as they engage in active or experiential learning are forms of assessment *as* learning. Here assessment *as* learning occurs when pre-service teachers are their own assessors. They monitor their own learning, ask questions, and use a range of strategies to decide what they know and can do, and how to use information for new learning (Chang 2018). In the context of the SLL, this includes pre-service teachers working out how to use fieldwork equipment and what data are best to collect, interpret, and analyze to answer the inquiry question. Additionally, working as part of a team to apply their skills and knowledge to real-world problems can be formatively assessed. These processes encourage criticality, reflexivity, and independent learning in line with the requirements of the curriculum document (Curriculum Planning & Development Division 2014).

6.5.2.2 Writing/Reflecting About the Fieldwork—Inquiry for Learning

In the SLL, pre-service teachers are encouraged to self-assess by reflecting on their observations, reactions, and feelings towards their surroundings. For example, pre-service teachers' observations are used to annotate photographs, allowing them to reflect on their personal experiences and thoughts. These reflections/annotations are collected and assessed by the instructor, who also identifies common or important points to be discussed during follow-up classes. Additionally, pre-service teachers peer assess by reading each other's work and providing comment (writing). Through self and peer assessment activities, a double loop of reflexivity for active learning is enabled (Chang 2018). Deep and reflective learning is inculcated (Brockbank and McGill 1998) by providing proper guidelines, objectives, and formative feedback for both students and teachers.

6.5.2.3 Relating the Formative Assessment Fieldwork Processes to the Product—Summative Presentation

The summative assessment task was the development of an inquiry-based fieldwork activity and presentation. By partaking in a geographical hands-on inquiry, pre-service teachers experienced the process as a learner and therefore could apply their reflections and feedback to select methodologies and approaches that would work effectively for the activity. In line with Looney (2011), this active and interactive learning process is counted as formative. In this way, pre-service teachers recognize what they need to think about for their future teaching such as what should be included to encourage students to recognize problems, ask questions, and make deeper connections that can be assessed at different stages of the inquiry process. Pre-service teachers will not only be able to formulate an optimal inquiry assessment but embed formative feedback to cultivate a double loop of reflexivity for both students and teachers, engaging in deeper experiential learning. Assessment *for* and *as* learning builds towards developing the knowledge and skills needed for the final product—the summative assessment piece.

6.6 Conclusion

This chapter has argued how inquiry can be used as a form of assessment in fieldwork. While traditional assessments in fieldwork lean towards standardized summative assessment tasks, the importance of ongoing formative assessments during fieldwork activities should not be ignored. Research has identified reflexivity as an important aspect of this process achieved by tools such as reflective diaries/journals. Such reflexivity resolves students' misconceptions, arouses students' interests in geographical issues, and develops higher order thinking skills such as criticality (Dummer et al. 2008). Feedback was also identified as another important aspect of assessment *for* and *as* inquiry, encouraging a double loop of reflexive pratice for both teachers and students (Brannon 2014). These important aspects were applied with pre-service teachers in an SSL in Singapore aligned with the GI approach from local curriculum guidelines. Pre-service teachers self and peer assessed whilst conducting the fieldwork activities. Feedback channels were witnessed in sharing and commenting on each others' work. They also annotated photographs—another form of reflexive practice—which promoted a double loop of reflexivity in their fieldwork assignments. All of these processes helped with the final product—the summative assessment task.

Ultimately, as educators, we desire effective assessment that will help improve student learning in geographical fieldwork. We believe that this can be achieved more effectively by viewing assessment as a process of inquiry throughout fieldwork. However, we need to go further. To take this field of research forward, fieldwork inquiry *for* and *as* assessment should also incorporate behavioral and emotional domains, providing opportunities for teachers and students to learn through assessment what it means to live sustainably in relation to the environment and other people.

References

Baidoo-Anu D, Owusu Ansah L, Acquah M, Bentil S (2019) Challenges teachers face in integrating fieldwork into teaching and learning of geography in senior high schools. Soc Sci Humanit J 3(7):1435–1444

Brannon S (2014) Assessment in fieldwork courses: what are we rating? J Educ Libr Inform Sci 55(4):274–302

Brockbank A, McGill I (1998) Facilitating reflective learning in higher education. Open University Press, Buckingham

Brown E (1969) The teaching of fieldwork and the integration of physical geography. In: Cooke R, Johnson J (eds) Trends in geography: an introductory survey. Pergamon Press, Oxford, pp 70–78

Butt G (2017) Debating the place of knowledge within geography education: reinstatement, reclamation or recovery? In: Brooks C, Butt G, Fargher M (eds) The power of geographical thinking. Springer, Cham, pp 13–26

Casinader N, Kidman G (2018) Fieldwork, sustainability, and environmental education: the centrality of geographical inquiry. Aust J Environ Educ 34(1):1–17. https://doi.org/10.1017/aee.2018.12

Chang C (2018) Enhancing geographical inquiry by integrating Google apps and NIE mGEO. Learing@NIE 2:18

Chang CH, Irvine K, Wu BS, Seow T (2018) Reflecting on field-based and technology-enabled learning in geography. In: Chang CH, Wu BS, Seow T, Irvine K (eds) Learning geography beyond the traditional classroom: examples from Peninsular Southeast Asia. Springer, Singapore, pp 201–212

Chang CH, Seow T (2018) Geographical education that matters—a critical discussion of consequential validity in assessment of school geography. Geogr Educ 31:31–40

Curriculum Planning & Development Division (2014) 2014 Lower secondary geography teaching syllabuses. Ministry of Education, Singapore

Curriculum Planning & Development Division (2021) Geography teaching and learning syllabuses: lower secondary express course normal (academic) course. Ministry of Education, Singapore

Das D, Chatterjea K (2018) Learning in the field—a conceptual approach to field-based learning in geography. In: Chang C, Wu B, Seow T, Irvine K (eds) Learning geography beyond the traditional classroom: examples from Peninsular Southeast Asia. Springer, Singapore, pp 11–34

Delandshere G (2002) Assessment as inquiry. Teachers Coll Rec 104(7):1461–1484. https://doi.org/10.1111/1467-9620.00210

Dummer T, Cook I, Parker S, Barrett G, Hull A (2008) Promoting and assessing 'deep learning' in geography fieldwork: an evaluation of reflective field diaries. J Geogr High Educ 32(3):459–479. https://doi.org/10.1080/03098260701728484

Ferretti J (2018) The enquiry approach in geography. In: Jones M, Lambert D (eds) Debates in geography education, 2nd edn. Routledge, Oxon, pp 115–126

Hill A, Dyment J (2016) Hopes and prospects for the sustainability cross-curriculum priority: provocations from a state-wide case study. Aust J Environ Educ 32(3):225–242. https://doi.org/10.1017/aee.2016.20

International Geographical Union—Commission on Geographical Education (2016) International Charter on Geographical Education. http://www.igu-cge.org/Charters-pdf/2016/IGU_2016_def.pdf. Accessed Aug 2016

Katz C (2009) Fieldwork. In: Gregory D, Johnston R, Pratte G, Watts MJ, Whatmore S (eds) The dictionary of human geography. Wiley-Blackwell, Chichester, pp 251–252

Kent MI, Gilbertson DD, Hunt CO (1997) Fieldwork in geography teaching: a critical review of literature of approaches. J Geogr High Educ 21(3):313–332

Kho E, Parker W (2010) Learning beyond the school walls: fieldwork in Singapore, grades 1–6. Soc Stud Young Learner 22(4):29–31

Lambert D, Reiss M (2014) The place of fieldwork in geography and science qualifications. Institute of Education, London, pp 1–19

Lim EQ (2018) The role of geographical investigations in developing students' cognitive thinking. HSSE Online 7(10):14–26

Looney JW (2011) Integrating formative and summative assessment: progress toward a seamless system? OECD Education Working Papers, No. 58. OECD Publishing. https://doi.org/10.1787/5kghx3kbl734-en

Meyer D, Antink Meyer A, Nabb K, Connell M, Avery L (2011) A theoretical and empirical exploration of intrinsic problems in designing inquiry activities. Res Sci Educ 43(1):57–76. https://doi.org/10.1007/s11165-011-9243-4

Moon J (2005) Guide for busy academics no. 4: learning through reflection. Higher Education Academy, York

Moore W (2003) Facts and assumptions of assessment: technology, the missing link. The J 30(6):20–26

Pedaste M, Mäeots M, Leijen Ä, Sarapuu S (2012) Improving students' inquiry skills through reflection and self-regulation scaffolds. Tech Instruct Cognit Learn 9:81–95

Pedaste M, Mäeots M, Siiman L, de Jong T, van Riesen S, Kamp E, Manoli CC, Zacharia ZC, Tsourlidaki E (2015) Phases of inquiry-based learning: definitions and the inquiry cycle. Educ Res Rev 14:47–61

Phillips R, Johns J (2012) Fieldwork for human geography. Sage Publications, London

Preston L (2016) Field 'work' vs 'feel' trip: approaches to out-of-class experiences in geography education. Geogr Educ 29:9–22

Rawling E (2007) Planning your key stage 3 geography curriculum. Geographical Association, Sheffield

Roberts M (2013) Geography through enquiry: approaches to teaching and learning in the secondary school. Geographical Association, Sheffield

Rogers E (2011) Fieldwork: assessment in outside—class environment. London Metropolitan Business School, London

Schön DA (1987) Educating the reflective practitioner: towards a new design for teaching and learning in the professions. Jossey Bass, San Francisco

Simms W, Shanahan M (2019) Using reflection to support environmental identity development in the classroom context. Environ Educ Res 25(10):1454–1478. https://doi.org/10.1080/13504622.2019.1574717

Summers M, Childs A, Corney G (2005) Education for sustainable development in initial teacher training: issues for interdisciplinary collaboration. Environ Educ Res 11:623–647

Wells G (1999) Dialogic inquiry: towards a sociocultural practice and theory of education. Cambridge University Press, New York

Chew-Hung Chang is a Professor with the Humanities and Social Science Education Academic Group of National Institute of Education, Nanyang Technological University. He is the Dean of Academic and Strategic Development at the National Institute of Education. As an academic, he has published extensively in geography and environmental education, in the areas of subject matter knowledge, curriculum, instruction, assessment, and teacher education.

Phoebe Ow is a Research Assistant at the National Institute of Education, Nanyang Technological University. Her research interests include climate change education, eco-governmentality, and sustainability developments and policy.

Part IV
Assessment as a Social Justice Endeavour in Geography Education

Chapter 7
The National Assessment of Educational Progress (NAEP): A Closer Look at US Gaps and Trends in Geography Achievement

Michael Solem

Abstract Administered by the National Center for Education Statistics within the US Department of Education, the National Assessment of Educational Progress (NAEP) programme provides statistically representative studies of what elementary and secondary students know and can do in different subject areas. NAEP results are reported as aggregated scores for student groups and as percentages of students that reach three achievement levels—Basic, Proficient, and Advanced. This chapter explores the performance of 8th grade (13–14 years old) students who participated in the 2018 NAEP geography assessment in relation to each achievement level. Examples of geography items at each level are provided to illustrate what geography students who perform at different levels are likely to know and understand. The chapter then proceeds to analyze the demographic and socioeconomic characteristics of students who score at a given NAEP achievement level. Results of NAEP geography assessments indicate no overall change in student achievement and persisting achievement gaps over a 25-year period. Implications of NAEP findings for research and practice in geography education are discussed in the context of curriculum theory advanced by the GeoCapabilities project and the results of other NAEP social studies assessments.

Keywords School geography · Student outcomes · Powerful knowledge

7.1 Introduction

Are young Americans 'proficient' in geography? Since 1994, answers to that question have been provided by the National Assessment of Educational Progress (NAEP). Colloquially known as 'The Nation's Report Card', NAEP is the largest continually operated assessment programme in the United States. Educators, policymakers, and

M. Solem (✉)
Texas State University, San Marcos, TX, USA
e-mail: msolem@txstate.edu

T. Bourke et al. (eds.), *Assessment in Geographical Education: An International Perspective*, Key Challenges in Geography,
https://doi.org/10.1007/978-3-030-95139-9_7

researchers value the programme for its deep descriptive reports about what elementary and secondary students know and can do in math, reading, writing, science, geography, civics, history, and other subjects that typically comprise the education of American students attending public and private schools.

To date, NAEP has assessed student achievement in geography five times: in 1994, 2001, 2010, 2014, and 2018. The first three assessments involved students at the 4th (9–10 years old), 8th (13–14 years old), and 12th (17–18 years old) grades. Beginning in 2014, only 8th grade students participated in the assessment. Because of factors ranging from budgetary cuts and a lack of geography education research and programmes that leverage NAEP data, geography was eliminated as an assessed subject area in 2019 by the National Assessment Governing Board (the Congressionally authorized council overseeing the NAEP programme) (Solem and Stoltman 2020). In all likelihood this means the nation will lose the ability to monitor and analyze gaps and trends in geography achievement until 2029, and perhaps beyond unless geographers start making enhanced uses of NAEP data to address important questions about educational policy and practice in the field.

In the spirit of advancing a research agenda involving NAEP geography data and reports, this chapter takes a closer view of the latest NAEP geography results for 8th graders. The chapter begins with a brief snapshot of the NAEP programme, describing how the assessment is conducted, and how results are reported. Examples of what students know and can do in geography are illustrated by a ‘map’ of released geography items on a scale with three cut off points set by NAEP to define achievement: Basic, Proficient, and Advanced. From there, the chapter probes into the characteristics of students who populate each performance level. The chapter concludes with a discussion of what the latest NAEP geography report suggests about the nation’s capacity to provide an opportunity for all students to learn geography. Future lines of research that leverage NAEP data for improving geographic learning outcomes are also discussed. This shows the value of large-scale assessment data and how it can be used to inform research, policy, and practice.

7.2 The NAEP Geography Assessment: An Overview

NAEP assessments are administered to students who are selected via a multistage, stratified systematic random sampling design that creates representative samples of the target population (e.g., all 4th graders in the US) (Brown et al. 2016). Like all trusted assessments, NAEP items are psychometrically designed to be valid and reliable measures of what children and young people know and are able to do in different school subjects. NAEP items are also subjected to a set of procedures known as differential item functioning for the purpose of eliminating test bias on the basis of race, ethnicity, and gender.[1]

[1] https://nces.ed.gov/nationsreportcard/tdw/analysis/scaling_checks_dif_proced.aspx.

The NAEP programme is conducted independently of assessments that are administered by states to measure achievement in different subjects. Because the US federal government has a fairly limited influence on the composition of school curricula, NAEP is not considered to be a 'high stakes' assessment—schools and teachers are not held accountable by state education agencies on the basis of how students perform on NAEP. Instead, NAEP provides a common measure of educational achievement that can be compared alongside the results of state-conducted assessments.

In geography, NAEP assesses student achievement in three content domains:

- *Space and place* questions measure students' knowledge of geography related to particular places on Earth, spatial patterns on the Earth's surface, and physical and human processes that shape spatial patterns.
- *Environment and society* questions measure students' knowledge of how people depend upon, adapt to, are affected by, and modify the natural environment.
- *Spatial dynamics and connections* questions measure students' ability to understand geography as it relates to spatial variations and the connections among people and places.

Within each content domain, questions test student cognition in three dimensions: knowing (what is it? where is it?); understanding (why is it there? how did it get there? what is its significance?); and applying (how can geographic knowledge and understanding solve problems?).

The NAEP geography assessment originated at the time national content standards for geography were first published in the volume *Geography for Life* (Geography Education Standards Project 1994). Although the NAEP geography assessment framework is organized differently, it is consistent with the disciplinary knowledge and skills recommended in the national standards (cf. Heffron and Downs 2012). NAEP's periodic reports on student achievement in geography are thus valued indicators of progress toward nationally recognized goals for geographic literacy (Solem and Stoltman 2020).

7.2.1 NAEP Geography Item Map

To aid the interpretation of NAEP results, NAEP publicly releases a set of assessment items and maps them onto a scale ranging from 0–300 to 0–500 (depending on the subject). NAEP scales represent the scores for students who were likely to answer a question correctly or to give a complete response. The NAEP item map for geography is based on a scale of 0–500 (Fig. 7.1).

NAEP item maps offer a glimpse into the assessed subject matter and the characteristics of items (e.g., multiple choice, constructed response) used to measure the extent a student demonstrates competency over that subject matter.[2] On NAEP item

[2] See the released items from the item map in the NAEP Questions Tool: https://nces.ed.gov/NationsReportCard/nqt/.

Scale from 0 – 500

500	
373	Describe positive and negative impact of Beijing growth—Complete (CR)
356	Analyze features in photograph for evidence of climate—Complete (CR)
342	Use temperature gradient map to identify and explain urban heat island—Complete (CR)
331	Identify cultural features immigrants bring to new places—Complete (CR)
328	Identify Cajun cultural region on map—Complete (SR)
315 ***NAEP Advanced***	
310	Locate and explain human interaction with natural hazards—Complete (CR)
308	Use time-zone map to determine date and time in a location—Complete (SR)
304	Identify difference between weather and climate—Complete (MC)
300	Describe positive or negative impact of Beijing growth—Partial (CR)
296	Use world map to identify a biome—Complete (MC)
287	Use map to identify benefit of summer monsoon to India—Complete (MC)
287	Identify land use from photo and explain its importance—Complete (CR)
282 ***NAEP Proficient***	
280	Use map to identify a rail hub and explain how a disruption can affect the network—Complete (CR)
276	Use map to relate congressional districts to population density—Complete (MC)
272	Locate and explain human interaction with a natural hazard—Acceptable (CR)
268	Analyze features in photograph for evidence of climate—Partial (CR)
265	Identify a cultural feature immigrants bring to new places—Partial (CR)
260	Identify that the tilt of Earth's axis affects the seasons—Complete (MC)
258	Identify terraced farming landscape from photograph—Complete (MC)
253	Identify crops grown in a tropical climate—Complete (MC)
244	Analyze satellite image of hurricane to select most destructive quadrant—Complete (SR)
242 ***NAEP Basic***	
237	Identify land use from photo and explain its importance—Partial (CR)
231	Use map to identify a rail hub or explain its importance in the larger network—Partial (CR)
201	Identify that heat from the sun causes water evaporation—Complete (MC)
182	Locate or explain human interaction with a natural hazard—Partial (CR)
0	

Fig. 7.1 NAEP item maps: Geography, 8th grade, 2018 (*Note* CR denotes a constructed-response question. SR denotes a selected-response question other than a single-select multiple-choice question. MC denotes a single-select multiple-choice question. *Source* US Department of Education, Institute of Education Sciences, National Center for Education Statistics, National Assessment of Educational Progress [NAEP], 2018 Geography Assessment)

maps, more difficult questions appear at the top and easier questions are located nearer the bottom. Constructed-response questions for which students could earn partial credit may appear on the map multiple times, once for each level of credit.

For example, only 8% of 8th graders participating in the 2018 assessment were able to give a complete response to an item measuring their ability to describe a positive and negative impact of urban growth in Beijing (Fig. 7.2). The location of this question on the map indicates that students with a score of 373 on the geography assessment had a high probability of answering this question fully and correctly.

NAEP item maps also include cut off points for the three NAEP achievement levels. These are defined in the NAEP geography assessment framework as follows (National Assessment Governing Board 2018, p. 44):

- *Basic* denotes partial mastery of the knowledge and thinking skills, but performance that is fundamental for adequate work in grades 4, 8, and 12.
- *Proficient* represents solid academic performance and competency over challenging subject matter. If students performed at the Proficient level on the

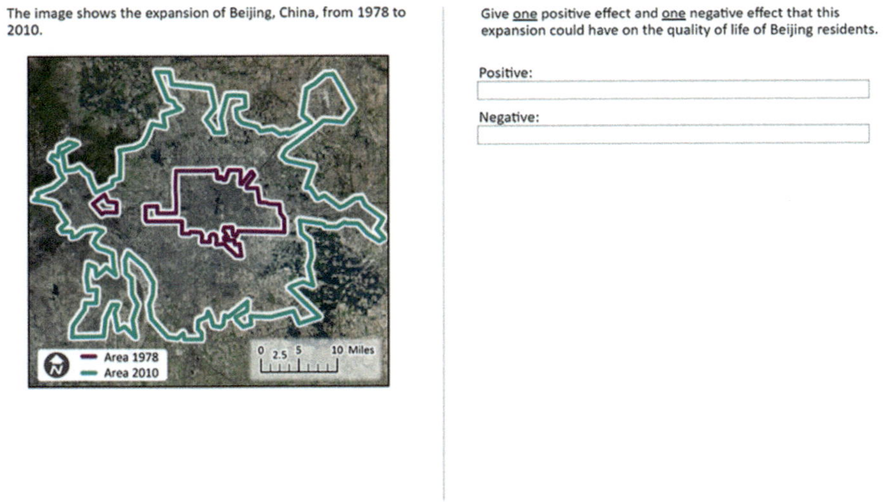

Fig. 7.2 Sample question for the geography subject NAEP 2018 assessment (*Source* US Department of Education, Institute of Education Sciences, National Center for Education Statistics, National Assessment of Educational Progress [NAEP], 2018 Geography Assessment)

assessment, the framework committees believe they would have learned enough geography to be competent students and productive citizens.

- *Advanced* performance on this assessment represents performance that is equal to that expected of top students in other industrialized nations—the ability to think critically about geographic issues and be able to integrate knowledge and skills into problem-solving situations.

The NAEP geography assessment framework document provides detailed descriptions of geography learning outcomes at each achievement level. The framework has remained the same since 1994 to permit trend analysis of student achievement (although new items appear in each assessment to measure achievement in relation to the framework).

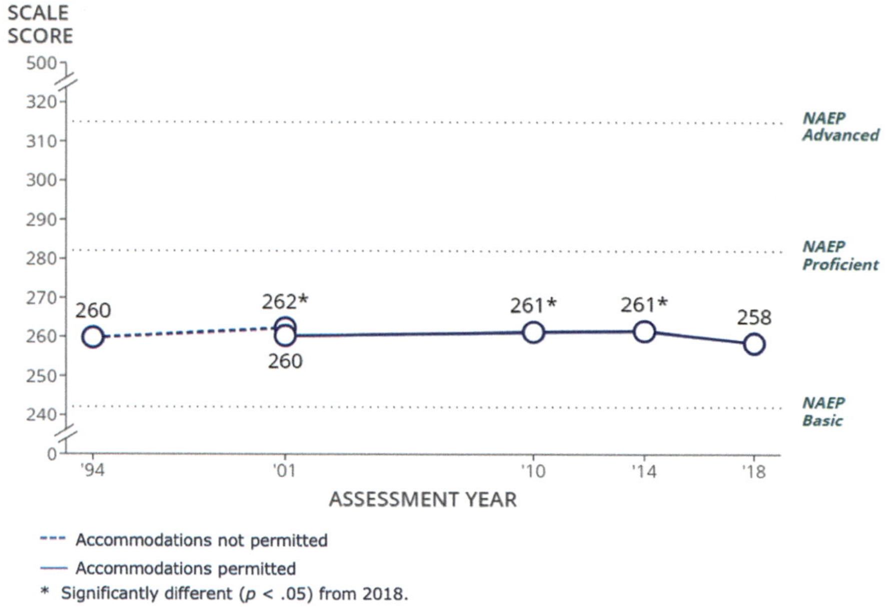

Fig. 7.3 Trend in 8th grade NAEP geography average scores (*Source* US Department of Education, Institute of Education Sciences, National Center for Education Statistics, National Assessment of Educational Progress [NAEP], 2018 Geography Assessment)

7.3 Comparing Student Characteristics at Different NAEP Achievement Levels

In 2018, a nationally representative sample of 12,900 8th grade students participated in the NAEP geography assessment. When the 2018 NAEP geography report[3] was released on April 23, 2020, some of the major findings included:

- There was no change in the 8th grade geography average score compared with 1994, which means three out of four 8th graders still perform below NAEP Proficient (Fig. 7.3);
- There was a statistically significant three-point decline in the geography average score since 2014; and
- 2018 scores are lower in the 'Space and Place' and 'Environment and Society' content areas compared with 2014; students also scored lower in 'Environment and Society' compared with all previous assessments in 2014, 2010, 2001, and 1994.

NAEP reports have also consistently documented achievement gaps in geography between different student groups. For example, since the first geography assessment in 1994, aggregate average geography scores for Hispanic (including Latino) students

[3] https://www.nationsreportcard.gov/highlights/geography/2018/.

Table 7.1 Percentages of students at each achievement level for 8th grade geography, by race/ethnicity in 2018

Year	Jurisdiction	Race/ethnicity	below Basic	at Basic	at Proficient	at Advanced
2018	National	White	16	49	32	4
		Black	58[c]	36[c]	5[c]	[a]
		Hispanic	41[c]	47	11[c]	1[c]
		Asian	17	44	34	5
		American Indian/Alaska Native	35	48	16	1
		Native Hawaiian/Other Pacific Islander	[b]	[b]	[b]	[b]
		Two or more races	25[c]	49	23[c]	3

[a]Rounds to zero
[b]Reporting standards not met
[c]Significantly different ($p < 0.05$) from White
Note Black includes African American, and Hispanic includes Latino. Race categories exclude Hispanic origin
Source US Department of Education, Institute of Education Sciences, National Center for Education Statistics, National Assessment of Educational Progress (NAEP), 2018 Geography Assessment

have barely surpassed the Basic performance level. Aggregate average geography scores for Black (including African American) students have never surpassed Basic.

The National Center for Education Statistics offers a tool, The NAEP Data Explorer,[4] to provide the public with a means to dig deeper into the results on any assessment. The following tables were generated with this tool to analyze the positioning of students on the NAEP achievement levels and how this varied on the basis of race and ethnicity, gender, and eligibility for the National School Lunch Program (see Sect. 7.3.3).

7.3.1 Race and Ethnicity

Table 7.1 shows the percentages of students in different racial and ethnic groups who score at the Basic, Proficient, and Advanced NAEP levels. Relative to White students, statistical t-tests indicate significantly higher percentages of Black, Hispanic, and multiracial students at the below Basic achievement level. At the Basic level, percentages of White students are significantly higher compared with their Black peers, whereas at the Proficient level, percentages of students who are White are significantly higher than for Black, Hispanic, and multiracial students. Only a very small

[4] https://www.nationsreportcard.gov/ndecore/landing.

Table 7.2 Percentages of students at each achievement level for 8th grade geography, by gender, 2018

Year	Jurisdiction	Gender	below Basic	at Basic	at Proficient	at Advanced
2018	National	Male	29	44	24	3
		Female	29	49[a]	20[a]	2[a]

[a]Significantly different ($p < 0.05$) from Male

Source US Department of Education, Institute of Education Sciences, National Center for Education Statistics, National Assessment of Educational Progress (NAEP), 2018 Geography Assessment

percentage of students performed at Advanced, with White students overrepresented when compared with Hispanic students. Statistically speaking, Black 8th graders are virtually nonexistent at the Advanced level.

7.3.2 Gender

Table 7.2 shows the percentages of male and female students who fall into the Basic, Proficient, and Advanced NAEP levels. *t*-tests of significance indicate no meaningful population differences at the below Basic level. The largest percentages of male and female students fall in the Basic category, with significantly higher percentages of females. Students in the Proficient and Advanced levels are more likely to be male than female.

7.3.3 National School Lunch Program Eligibility

The US government, through the National School Lunch Program (NSLP), provides subsidized or free lunches to students attending public and nonprofit private schools and residential child care facilities. To be eligible, a student's household income must fall below a certain percentage of the federal poverty level or qualify via participation in various federal assistance programmes, such as Head Start and the Supplemental Nutrition Assistance Program. Homeless, migrant, runaway, and foster children also qualify. NSLP is thus a commonly used indicator for comparing students on the basis of socioeconomic status.

Table 7.3 shows the percentages of NSLP eligible and ineligible students who perform at the Basic, Proficient, and Advanced NAEP levels. Percentages of students who are ineligible for free/reduced lunch are significantly lower at the below Basic level, and significantly higher at the Basic, Proficient, and Advanced levels, compared with students who are eligible.

What factors related to curriculum and instruction might potentially explain these patterns in geography achievement? A prior analysis of student achievement using

Table 7.3 Percentages at each achievement level for 8th grade geography, by National School Lunch Program eligibility, 2018

Eligible	Not eligible	Information not available
below Basic	below Basic	below Basic
45	14<	16<
at Basic	at Basic	at Basic
44	49>	46
at Proficient	at Proficient	at Proficient
11	33>	33>
at Advanced	at Advanced	at Advanced
#	4>	5>

Rounds to zero
< Significantly lower than Eligible
> Significantly higher than Eligible
Source US Department of Education, Institute of Education Sciences, National Center for Education Statistics, National Assessment of Educational Progress (NAEP), 2018 Geography Assessment

the NAEP Data Explorer identified some curriculum and instruction-related factors that were found to be inversely related to student scores on the geography assessment (Solem 2021):

> Students who indicated no engagement with inquiry and problem solving scored significantly higher compared with their peers who 'sometimes' or 'always or almost always' learned geography through these approaches. Similarly, there were significant differences in geography achievement scores when comparing students on the basis of how frequently they participated in geography field trips and class debates and discussions about the environment: the more regularly students experienced these pedagogies, the worse they apparently fared on the assessment.

To explore this issue further, the following tables were produced with the NAEP Data Explorer to analyze the positioning of students on NAEP achievement levels who receive different amounts of exposure to these active pedagogies. Each table compares percentages of students at different NAEP performance levels relative to students who responded 'never' to the question posed in the item.

Table 7.4 reports percentages of students at each NAEP achievement level on the basis of how frequently they use geographic information to answer questions or solve problems. Compared with students who responded 'never or hardly ever',

- Significantly higher percentages of students whose responses were 'sometimes', 'often', or 'always or almost always' are found in the below Basic category.
- Significantly lower percentages of students whose responses were 'sometimes' and 'always or almost always' are found at Proficient.
- No significant differences were found in the percentages of students at Basic and at Advanced.

Table 7.5 reports percentages of students at each NAEP achievement level on the basis of how frequently they participate in field trips about geography topics. Results

Table 7.4 Percentages of students at each achievement level for 8th grade geography, by 'Use geographic information to answer questions or solve problems', 2018

Percentage	below Basic	at Basic	at Proficient	at Advanced
Never or hardly ever				
27	24	48	25	3
Once in a while				
22[a]	26	47	25	3
Sometimes				
25[a]	31[a]	46	21[a]	2
Often				
16[a]	28[a]	47	22	3
Always or almost always				
9[a]	38[a]	44	16[a]	2

[a]Significantly different ($p < 0.05$) from Never or hardly ever
Note Some apparent differences between estimates may not be statistically significant
Source US Department of Education, Institute of Education Sciences, National Center for Education Statistics, National Assessment of Educational Progress (NAEP), 2018 Geography Assessment

Table 7.5 Percentages of students at each achievement level for 8th grade geography, by 'Attend field trips about geography topics'

Percentage	below Basic	at Basic	at Proficient	at Advanced
Never				
73	22	49	26	3
Once				
15[a]	39[a]	44[a]	16[a]	2
2–3 times				
8[a]	42[a]	44	13[a]	1[a]
4–5 times				
2[a]	67[a]	27[a]	6[a]	[b]
More than 5 times				
2[a]	78[a]	18[a]	3[a]	[b]

[a]Significantly different ($p < 0.05$) from Never
[b]Rounds to zero
Source US Department of Education, Institute of Education Sciences, National Center for Education Statistics, National Assessment of Educational Progress (NAEP), 2018 Geography Assessment

indicate that nearly three out of four 8th graders did not attend a field trip in 2018. This is unsurprising given the cost and liability concerns that frequently surround field trips. Still, the analysis revealed significant differences on the basis of frequency of field trip attendance. Compared with students whose response was 'never',

Table 7.6 Percentages of students at each achievement level for 8th grade geography, by 'Take part in environmental debates or panel discussions'

Percentage	below Basic	at Basic	at Proficient	at Advanced
Never				
59	23	48	25	3
Once				
18[a]	30[a]	46	21[a]	3
2–3 times				
13[a]	35[a]	45	18[a]	1
4–5 times				
5[a]	45[a]	39[a]	15[a]	1
More than 5 times				
4[a]	45[a]	37[a]	17[a]	2

[a]Significantly different ($p < 0.05$) from Never
Rounds to zero
Source US Department of Education, Institute of Education Sciences, National Center for Education Statistics, National Assessment of Educational Progress (NAEP), 2018 Geography Assessment

- There are statistically significant and progressively higher percentages of students at the below Basic level with each incremental increase in field trip attendance.
- There are statistically significant and progressively lower percentages of students at the Proficient level with each incremental increase in field trip attendance.

Table 7.6 reports percentages of students at each NAEP achievement level on the basis of how frequently they take part in environmental debates or panel discussions. Results indicate that over half of 8th graders did not experience this classroom activity, with the remaining students spread over the other response categories. Compared with students whose response was 'never',

- There are significantly higher percentages of students at the below Basic level who responded 'once', '2–3 times', '4–5 times', and 'more than 5 times'.
- There are significantly lower percentages of students at the Basic level who responded '4–5 times' and 'more than 5 times'.
- There are significantly lower percentages of students at the Proficient level who responded 'once', '2–3 times', '4–5 times', and 'more than 5 times'.

It appears from Tables 7.4–7.6 that the more often students are exposed to the three types of active pedagogies, the more likely they are to score at lower levels on the NAEP achievement scale. These findings, while perhaps surprising, are consistent with a multilevel statistical analysis of the NAEP US history assessment. In that study, the authors found incremental increases in exposure to so-called multimodal instruction (e.g., frequency of group projects, going on field trips, classroom presentations, writing reports) were linked to statistically significant decreases in students' knowledge of history as assessed by NAEP (Fitchett et al. 2017).

However, further research is needed to dig deeper into the characteristics of students who are grouped together within each of the curriculum and instruction response categories reported in Tables 7.4–7.6. Each response (e.g., 'never', 'often') groups students together who are otherwise diverse in terms of their background characteristics, classrooms, and schools. Until research emerges that statistically controls for the 'effects' of these covarying independent variables, it is important to interpret the data with caution and avoid jumping to conclusions about the independent effects of these active pedagogies on geography achievement.

7.4 Discussion

NAEP geography is a national-level assessment, but the US does not have a uniform geography curriculum. Furthermore, there are many challenges in providing K–12 geography instruction within and between the states (Government Accountability Office [GAO] 2015). Foremost among them is a pressure to teach other subjects, most notably reading and math that many states are accountable for assessing intensively to comply with funding requirements. Additional challenges include low levels of support for geography education among stakeholders including policymakers, public misunderstanding of the nature of geography, uneven access in schools to high quality instructional materials and technologies, and low numbers of teachers with an academic background in geography. All of these challenges potentially restrict access to geography in schools in ways that are inequitable and place certain student groups at an educational disadvantage.

Further research is needed to explore the extent that the national outcomes reported in this chapter can be explained by differences in curriculum content and exposure, instructional methods, classroom resources, and teacher characteristics. NAEP itself can be further leveraged by researchers to explore the relationship between geography achievement and classroom-level variables of this nature. This is possible because NAEP conducts extensive background surveys with students, teachers, and school principals to gather contextual data alongside the assessments themselves. Data from these surveys are available in the NAEP Data Explorer, but researchers can also request restricted data (individual level responses) to develop predictive models of geography achievement. Research in social studies education involving restricted NAEP data provides geographers with examples of how achievement can be studied in relation to factors such as teaching time, teacher training, parental income, and many other contextual classroom-level factors (e.g., Fitchett et al. 2017; Fitchett and Heafner 2017; Heafner and Fitchett 2015; Heafner et al. 2019).

7.4.1 Geography Achievement Disparities as Capabilities Deprivation

The general lack of public support and understanding of geography may well lead some to view the NAEP geography assessment and ask, why does it matter? To grasp the significance of NAEP reports about geography achievement, we need to appreciate what it is that geography *adds* to the education of young people. It is here where recent work in GeoCapabilities provides a productive response.

GeoCapabilities is an international collaborative project that argues for a geography curriculum based upon conceptions of powerful knowledge (Lambert et al. 2015; Lambert and Solem 2017; Biddulph et al. 2020). Powerful knowledge is a complex concept, and its origins and applications in education have been extensively and critically discussed (Maude 2016; Huckle 2019; Lambert 2019; Virranmäki et al. 2019; Young 2008). In the simplest terms, powerful knowledge refers to disciplinary knowledge that enables young people to think in ways that expand their freedoms and opportunities (capabilities) to achieve well-being (Nussbaum and Sen 1993; Nussbaum 2011). Unlike 'everyday' knowledge that individuals acquire from their interactions and experiences in the world, powerful knowledge is often abstract and conceptual, and at times it is contested. A curriculum based on powerful knowledge 'does not lead to a list of content that must be taught, but only to ways of thinking that should be developed through whatever content is selected' (Maude 2016, p. 75). Curriculum making by teachers thus becomes the crucial ingredient in developing the capacity of young people to think geographically.

This forces a crucial question: what do the outcomes of the NAEP geography assessment imply about the nature of school geography? Recall that NAEP is a measure of disciplinary knowledge and skills inclusive of physical geography, human geography, and environmental geography. From the perspective of GeoCapabilities, two observations can be made. First, the persisting gaps in geography achievement represent a form of 'capabilities deprivation', that is to say a systemic denial of the 'pedagogic rights' of all young people to access powerful knowledge that is crucial for individual enhancement, social inclusion, and political participation (Bernstein 2000). Confronting the inequities and disparities in school geography outcomes should therefore be seen as a social justice imperative (cf. Biddulph et al. 2020).

A second concern is the apparent disconnect between the geography measured by NAEP and the actual learning experiences of young people in schools. The GeoCapabilities project warned of the rise of overly-socialized 'progressive' curricula, in which 'knowledge is seen as a mere arbitrary or neutral vehicle for engaging in supposedly more valued generic skills and competencies'—primarily through active and multimodal forms of pedagogy (Bustin et al. 2020, p. 202). It is true that instructional practices such as inquiry, fieldwork, and issue-oriented classroom debates have a long tradition in geography education and there is plenty of research to show these methods can 'work' to improve learning outcomes and interest in the subject (Klein et al. 2019). At the same time, care must be taken by teachers in their curriculum making to ensure their pedagogies result in the acquisition of disciplinary knowledge and skills. While students may well gain useful skills from 'content agnostic' models of inquiry and fieldwork such as National Geographic's Geo-Inquiry Process (Oberle 2020), the trade-off may be less time spent learning disciplinary knowledge and skills that are significant for human capability development.

The 'curriculum of engagement' in powerful knowledge that GeoCapabilities envisions is compatible with active pedagogies, but also concerned with the epistemic quality of what students learn. It is an idea that continues to spread internationally and inspire productive discussions about the purpose and significance of geography in education (Bustin et al. 2020). Given what we know thus far from studies based on the NAEP geography assessment, future research should probe more deeply into factors that may be undermining the epistemic quality of the geography curriculum in ways that adversely affect the capabilities of girls, students of color, and students from less privileged social and economic circumstances.

7.5 Conclusion

After five NAEP geography assessments over 25 years, this much is clear: schools in the US have not only made no progress in bolstering overall achievement levels in geography, but they simultaneously have failed to close achievement gaps in the subject. This is true despite the fact that the same period (1994–2018) featured major attempts to reform geography in the nation's schools, from two editions of national standards to related state-based teacher training institutes sponsored by the National Geographic Alliance Network.

The future growth and character of the academic geography discipline may well depend on making broad-scale improvements in geography achievement at the school level. Of the approximately 2.5 million students who typically enter college each year in the US, only 4,000–5,000 can be expected to graduate with an undergraduate degree in geography (American Association of Geographers [AAG] 2019). Those geography majors are disproportionately White and male relative to other social and environmental sciences (Keen 2018a, b). The geography discipline has changed very little demographically over the same period of the NAEP geography assessment.

Although NAEP has repeatedly reminded us of the major challenges in offering a high quality and equitable school education in geography, it also offers a path toward a better future. Geographers can leverage NAEP's tools and data to address research questions that can impact policy and practice recommendations by professional associations that NAEP data are uniquely positioned to address. Research for improving national learning outcomes in geography will require careful coordination and leadership among geography organizations, working in partnership with researchers, practitioners, and policymakers who can conceive and implement practices that evidence suggests can ensure the next generation of American youth encounter geography's powerful knowledge and are motivated to learn and use that knowledge to achieve their potential in life, work, and citizenship.

References

American Association of Geographers [AAG] (2019) Guide to geography programs in the Americas. AAG, Washington, DC

Bernstein B (2000) Pedagogy, symbolic control and identity. Rowman and Littlefield, Oxford

Biddulph M, Bèneker T, Mitchell D, Hanus M, Leininger-Frézal C, Zwartjes L, Donert K (2020) Teaching powerful geographic knowledge—a matter of social justice: initial findings from the GeoCapabilities 3 project. Int Res Geogr Environ Educ. https://doi.org/10.1080/10382046.2020.1749756

Brown NJS, Dai S, Svetina D (2016) Analyzing NAEP data at the item level. In: Kloosterman P, Mohr D, Walcott C (eds) What mathematics do students know and how is that knowledge changing? evidence from the National Assessment of Educational Progress. Information Age, Charlotte, pp 33–44

Bustin R, Lambert D, Tani S (2020) The development of GeoCapabilities: reflections, and the spread of an idea. Int Res Geogr Environ Educ 29(3):201–205. https://doi.org/10.1080/10382046.2020.1749773

Fitchett PG, Heafner TL (2017) Student demographics and teacher characteristics as predictors of elementary-age students' history knowledge: implications for teacher education and practice. Teach Teach Educ 67:79–92

Fitchett PG, Heafner TL, Lambert RG (2017) An analysis of predictors of history content knowledge: implications for policy and practice. Educ Pol Anal Arch 25(65). https://doi.org/10.14507/epaa.25.2761

Geography Education Standards Project (1994) Geography for life: national geography standards. National Geographic Society Committee on Research and Exploration, Washington, DC

Government Accountability Office [GAO] (2015) K–12 education: most eighth grade students are not proficient in geography (GAO-16-7). United States Government Accountability Office, Washington, DC

Heafner TL, Fitchett PG (2015) An opportunity to learn US history: what NAEP data suggest regarding the opportunity gap. High Sch J 98(3):226–249

Heafner TL, VanFossen PJ, Fitchett PG (2019) Predictors of students' achievement on NAEP-economics: a multilevel model. J Soc Stud Res 43(4):327–341

Heffron SG, Downs RM (eds) (2012) Geography for life: National Geography Standards, 2nd edn. National Council for Geographic Education, Washington, DC

Huckle J (2019) Powerful geographical knowledge is critical knowledge underpinned by critical realism. Int Res Geogr Environ Educt 28(1):70–84

Keen J (2018a). Gender in geography. AAG, Washington, DC. www.aag.org/disciplinarydata. Accessed 25 May 2020

Keen J (2018b) Race and ethnicity in geography. AAG, Washington, DC. www.aag.org/disciplinarydata. Accessed 25 May 2020

Klein P, Barton K, Salo J, Lee J, Vowles T (2019) Conveying geographic concepts through issues-based inquiry. In: Walkington H, Hill J, Dyer S (eds) Handbook of teaching and learning in geography. Edward Elgar Publishing, Cheltenham, pp 211–226

Lambert D (2019) On the knotty question of 'recontextualising' geography. Int Res Geogr Environ Educ 28(4):257–261

Lambert D, Solem M (2017) Rediscovering the teaching of geography. Geogr Educ 30:8–15

Lambert D, Solem M, Tani S (2015) Achieving human potential through geography education: a capabilities approach to curriculum making in schools. Ann Assoc Am Geogr 105(4):723–735

Maude A (2016) What might powerful geographical knowledge look like? Geogr 101:70–76

National Assessment Governing Board (2018) Geography Assessment Framework for the 2018 National Assessment of Educational Progress. National Assessment Governing Board, Washington, DC

Nussbaum M (2011) Creating capabilities: the human development approach. Harvard University Press, Cambridge

Nussbaum M, Sen A (eds) (1993) The quality of life. Clarendon Press, Oxford

Oberle A (2020) Advancing students' abilities through the geo-inquiry process. J Geogr 119(2):43–54. https://doi.org/10.1080/00221341.2019.1698641

Solem M (2021) The NAEP Data Explorer: digging deeper into K–12 geography achievement and why it matters. Geogr Teach. https://doi.org/10.1080/19338341.2021.1895864

Solem M, Stoltman J (2020) The NAEP geography assessment: will it come back? J Geogr 119(4):136–139. https://doi.org//https://doi.org/10.1080/00221341.2020.1773903

Virranmäki E, Valta-Hulkkonen K, Rusanen J (2019) Powerful knowledge and the significance of teaching geography for in-service upper secondary teachers—a case study from Northern Finland. Int Res Geogr Environ Educ 28(2):103–117

Young M (2008) Bringing knowledge back in. Routledge, London

Michael Solem is Professor of Geography at Texas State University. After 14 years directing educational programs on staff at the American Association of Geographers, he continues to serve the AAG as Senior Adviser for Geography Education and Co-Director of the National Center for Research in Geography Education. Michael is co-chair of the International Geographical Union's Commission on Geographical Education (2020–2024). He is the 2015 recipient of the AAG Gilbert Grosvenor Honors in Geographic Education.

Chapter 8
Focusing on Quality, Forgetting Inequalities: Assessment Within GIS in the Geography Curriculum and Assessment Policy Statement (CAPS) in South Africa

Sadhana Manik

Abstract After becoming a democratic country in 1994, South Africa set a course to overhaul all aspects of society. Education had a transformational agenda to overcome inequalities: first, removing apartheid influences, and second, in being responsive to local requirements of an equal education for a diverse population. To achieve quality education, school curricula were transformed from the introduction of an initial democratic curriculum in 1997 to the recent Curriculum and Assessment Policy Statement (CAPS) which forms part of the National Curriculum Statement. Preparing learners to acquire skills to be part of a rapidly advancing technology-driven world has led to recent changes. In high school geography, the CAPS document includes the theme Geographical Information Systems (GIS) from Grades 10 to 12 and is tested in the Grade 12 exit examinations. While previous work has critiqued an earlier geography curriculum using 'continuity' and 'progression' as markers of quality education, the geography CAPS has not been analyzed using these concepts. Hence, in this chapter I analyze the assessments in GIS, contending that some quality markers which articulate continuity, sequencing, and progression in assessments are evident. However, despite attention to quality, historical inequalities in public schools linger, subverting the intended curriculum. Learners in disadvantaged schools where resources (teacher training and tools) are absent/inadequate are taught 'about GIS' and not 'through GIS', undermining teaching, learning, and assessment in GIS.

Keywords South Africa · Quality · Inequalities · Curriculum · Assessment · Geographic Information Systems

S. Manik (✉)
University of KwaZulu-Natal, Durban, South Africa
e-mail: manik@ukzn.ac.za

T. Bourke et al. (eds.), *Assessment in Geographical Education: An International Perspective*, Key Challenges in Geography,
https://doi.org/10.1007/978-3-030-95139-9_8

8.1 Introduction: An Overview of Changes in Education and Geography Education in South Africa

Inequality on the basis of race permeated all aspects of life in apartheid South Africa, creating divisions between Whites and Blacks with unfair treatment as the basic principle. Education, where funding was distributed per child on the basis of South African racial categories, was responsible for preparing the majority non-White population to accept mediocrity in society. For example, Bantu Education (see the Bantu Education Act of 1953; Union of South Africa 1953), with an inferior curriculum, was specifically carved out for the African population (Nel and Binns 1999). During this time geography, like other school subjects, was prone to particular circumstances such as large class sizes, inadequate instructional materials (Eiselen 1953), and insufficiently qualified teachers in Black schools. Although Christian national education (which was steeped in authoritarianism) was the preserve of Whites (Manik 2016), there were few similarities in teaching and learning practices in schools. For example, teacher-centred lessons, rote learning (Nel and Binns 1999), and summative assessment strategies were the hallmarks of geography lessons for the majority of practicing geography teachers during apartheid in South Africa (1948–1994). During this time, assessment of learning dominated in the form of quarterly examinations at the end of each of the four school terms (punctuated with some tests in between). These types of examinations were the norm in public schools for non-Whites. The idea was 'to create a working class at minimum cost to the apartheid state' (Manik 2016, p. 446) but as Fleisch (2002) argues, Bantu Education did not aim to necessarily mold people into being submissive. Rather, the motive was political; the African population were being prepared for certain sectors of employment in apartheid society whilst the White population were ensured socio-economic and political dominance legitimizing their supremacy in South Africa.

After becoming a democratic country in 1994, South Africa set a course to overhaul all aspects of society. A first step in moving towards equality was in dissolving the 19 departments of education, collapsing them into nine provincial departments with a national education department (Christie 2008). The White Paper on Education and Training (Department of Education 1995) spelt out the transformation of education that was envisaged (Lentsa 2019). Key principles of democracy, equality, and redress led to the National Education Policy Act of 1996 (NEPA) with specific statements on quality, language, curriculum, and assessment. Education, therefore, had a specific transformational agenda to remove apartheid influences and continually work towards achieving social justice whilst building an equal educationally sound foundation for all learners. Education was mooted as being responsive to local socio-economic and political needs and the requirements of a diverse population hungry for educational opportunities and positive post-apartheid change. Despite this, critique was leveled by esteemed education scholars.

Instead of real improvements in equality and quality in education, changes were a simple box ticking exercise 'superficially cleansing' (Jansen 1999) the apartheid school curriculum. Lentsa (2019), drawing on writers such as Motala (2011) and

Sayed (2002), stated that 'ongoing policy goals and initiatives to improve access to quality basic education have not been successful' (p. 7). For example, Spaull (2013), citing the Southern and East African Consortium for Monitoring Educational Quality (SACMEQ) reports, revealed shockingly low numeracy and literacy skills of learners from Grades 1–12. More recently, Lentsa (2019) has agreed with these earlier sentiments, stating that 'quality educational outcomes such as literacy and numeracy skills are inadequate, especially among poor and rural learners' (p. 7) and in South Africa, rural poor learners are Black African. Jansen and Christie (1999) claimed that early education curricular iterations were purely symbolic and rooted in political ideology. Jansen (2002) in later work stated that 'despite unprecedented investment in policymaking and policy production in the years anticipating and following the end of legal apartheid in South Africa, there appears to be very little change in the daily routines of schools and classrooms of the nation' (p. 199). Lentsa (2019, p. 8) concluded that 'there is a consensus among scholars [that] there are major implementation gaps in the policy practice'. Obviously, such gaps will derail any efforts to ensure quality and equal education for all learners in public schools in South Africa.

In terms of geography education, commentators such as Wilmot and Dube (2015) argued that change processes were 'complex and fraught with difficulties and mis-assumptions' (p. 95). Quality education in a globalized world is also about responsiveness to change. The arrival of what is being termed the fourth industrial revolution is an important paradigm shift that impacts on education curricula with the need for specific high skills acquisition. The fourth industrial revolution is hailed to re-configure the world given advancements in technology (for example, robotics, nanotechnology, artificial intelligence, etc. which are closely connected to Geographical Information Systems [GIS]) and thus education systems are adopting emerging technologies to be relevant to these rapidly changing times (Schwab 2017). This adds a new caveat on policy implementation in the quality and inequalities discourse in South African public schools.

8.2 Selected Curriculum Changes Prior to CAPS

In South Africa schooling is currently separated into two phases: General Education and Training (GET) phase (Reception [R] to Grade 9) and Further Education and Training (FET) phase (Grades 10–12). The FET phase marks the final three years of high school, providing a career focused education. In setting the landscape for a discussion and analysis of assessment continuity, sequencing, and progression in the theme of GIS in the Curriculum and Assessment Policy Statements (CAPS) currently in use, it is useful to briefly outline literature related to earlier curriculum changes in education and geography education.

School education has undergone numerous reforms, so much so that Wilmot and Dube (2015, p. 95) aptly dubbed it 'policy overload'. The most significant of these changes (prior to the CAPS launched in January 2012), aimed to strengthen the quality of education in South Africa, included: an interim syllabus (1996), followed by Curriculum 2005 (C2005) (1997), and then the Revised National Curriculum Statement (2002). Geography education curriculum scholars Beets and Le Grange (2008, p. 68) have argued that 'these changes have ostensibly been introduced to strengthen social sciences education in South Africa generally, and geography education more specifically'. Key changes in the new curricula were a move away from a content-centred curriculum to an outcomes-based one (with clearly defined assessment criteria) and integration of knowledge through the merging of certain subjects. For example, history and geography were combined in the GET phase of C2005 and called Human and Social Sciences (HSS), but in the FET phase they remained in separate disciplines. Whilst scholars such as Van Harmelen (1999) praised C2005 for moving away from textbook-based teaching, Wilmot and Dube (2015) commented on how two decades of policy change to improve and streamline curriculum had negatively impacted teachers who felt pedagogically overwhelmed. Indeed, one of the greatest critiques levelled against C2005 (Jansen and Christie 1999) related to whether teachers in poorly resourced (disadvantaged) schools could implement the curriculum. This prompted the then Minister of Education to initiate a review (see Chisholm et al. 2000). The revised national curriculum statement (RNCS) followed, which retained outcomes-based education and introduced a simplified curriculum with learning outcomes and assessment standards. Le Grange and Beets (2005) specifically critiqued C2005 and the RNCS using the concepts of 'continuity' and 'progression' as markers for quality education. They noted shortcomings in the assessment frameworks but asserted that 'attempts have been made to ensure that *continuity* and *progression* are elements of post-apartheid geography education through the inclusion of design features such as range statements and assessment criteria' (Beets and Le Grange 2008, p. 68).

8.3 Curriculum Review and a Combined NCS

In 2009, there was a further review of the curricula (the RNCS for Grades R–9 and the NCS for Grades 10–12). This review led to the construction of a combined National Curriculum Statement (NCS) for Grades R–12 launched in January 2012 (which is the current education policy document) and it updates the previous curricula and provides clarity on the topics to be taught in each of the four school terms in a year. This NCS comprises three documents:

- Curriculum and Assessment Policy Statements (CAPS) (Department of Education 2012) for each school subject,
- A National Policy on Program and Promotion Requirements of the NCS, and
- The National Protocol for Assessment Grades R–12.

  The NCS was progressively introduced in schools with Grade 12 experiencing the new curriculum for the first time in 2014. The NCS is based on the following seven principles (Department of Basic Education 2011):

- social transformation,
- active and critical learning,
- high knowledge and high skills,
- progression,
- human rights, inclusivity, environmental and social justice,
- valuing Indigenous knowledge systems, and
- credibility, quality, and efficiency in education.

It is valuable that quality and equality in education is foregrounded. Additionally, active and critical learning and high knowledge and skills are highlighted. These principles have relevance for the majority Black African learners in disadvantaged schools especially in respect to how GIS is taught and the nature of the assessments in the FET phase.

Having presented the backdrop to curriculum development in South Africa, the chapter is further divided into four sections. First the CAPS document is outlined and the nature of assessments, bringing us to the present curriculum/assessment moment. Second, the concepts of continuity, sequencing, and progression are explained. Then, these concepts are applied to the theme of GIS in the CAPS document with a specific focus on assessment. Next, an analysis of the GIS assessments and the public schooling context are presented to show areas of disconnect. The chapter culminates with a discussion of significant threads which indicate that there are deeply rooted inequalities in education which remain from the apartheid era and their contextual realities undermine the successful implementation of GIS in public schools in South Africa.

8.4 Present Curriculum and Assessment Moment

In the FET phase from Grades 10–12, a learner takes seven subjects of which four are mandatory and three are the learner's choice from a list of 25 subjects (geography is one such subject). The instructional time for Grades 10–12 is a total of 27.5 h per week. The home language, first additional language, and mathematics (all compulsory subjects) have the greatest time allocation of 4.5 h per week. Life Orientation is also a compulsory subject allocated two hours per week. Each of the optional subjects, like geography, are allocated four hours per week.

The CAPS for geography (Department of Basic Education 2011) specifies the learning, teaching, and assessments for each grade of schooling focused on the achievement of knowledge, skills, and values. The key questions geographers ask, content topics, and allocated times, as well as a list of additional resources, are outlined. The basic resources (such as the tools to teach) to meet all the curriculum requirements are not mentioned. The weighting of each of the topics and sub-topics is clearly stated. In addition, each term is dedicated to a different focus in geography. The grades and topics per term are captured in Table 8.1 (Department of Basic Education 2011, p. 12).

Table 8.1 Grades and topics in the FET phase (Department of Basic Education 2011)

Grade and term	Topic
10	Geographic skills and techniques
1	The composition and structure of the Earth
2	Plate tectonics, folding, faulting, volcanoes, earthquakes
3	Population: structure, growth and movement
4	Water resources: water in the world: oceans, flooding, management
11	Geographic skills and techniques
1	Global air circulation, Africa's weather and climate
2	Rocks and landforms, slopes, mass movements
3	Development: differences, issues and opportunities
4	Resources and sustainability: soil, energy
12	Geographic skills and techniques
1	Climate and weather: cyclones, local climate
2	Geomorphology: drainage systems and fluvial processes
3	Rural and urban settlement
4	Economic geography of South Africa

The assessment section in the CAPS document emphasizes assessment as 'a continuous, planned process' (Department of Basic Education 2011, p. 49). The critical link between teaching, learning, and assessment is foregrounded with the necessity for assessment highlighted as significant for both the teacher and learner. The CAPS document draws attention to the National Protocol on Assessment (Department of Education 2005) as well as the Geography Subject Assessment guidelines which must be read in tandem with the CAPS for clarity on the requirements per grade in the FET phase.

The CAPS foregrounds both formative and summative assessment as necessary in the FET phase in geography. The CAPS further advocates when these assessments should be used, implying that there should be a continuous series of assessments that unfold in the geography classroom to develop the learner: 'both assessment for learning and assessment of learning strategies should be used during the school year' (Department of Basic Education 2011, p. 49). Assessment *for* learning is explained as 'informal assessment' and is perceived to be a 'daily assessment' (Department of Basic Education 2011, p. 49). Further, 'This type of assessment is done, for example, through observations, in discussions, during practical demonstrations, in learner/teacher conferences, and in informal classroom interactions' (Department of Basic Education 2011, p. 49). The CAPS document also asserts that formal assessment in the format of tests, examinations, and practical tasks should be undertaken. Interestingly, formal assessments are the only tasks used for certification as they are subjected to moderation to assure quality so that 'appropriate standards can be maintained' (Department of Basic Education 2011, p. 49).

The format of the formal assessments and their weightings are captured in Table 8.2. As shown, there are numerous assessments in Grades 10 and 11 which only account for 25% of the final weighting. In contrast, the end of the year examination made up of two tests accounts for 75% of the marks. In Grade 12, the three specified tasks are the same; however, without GIS application. There is also the addition of a trial examination.

Table 8.3 further details the weightings for GIS in the geography examinations in the FET phase.

Having outlined in detail the topics in the curriculum, the types of assessment, their weightings, and how GIS is represented, I now turn my attention to defining the concepts (namely continuity, sequencing, and progression) which will be used to analyze the GIS theme in the CAPS document.

Table 8.2 Weighting of assessments in the FET phase (Department of Basic Education 2011)

Grade	Formal assessment	Weighting
10 and 11	Task 1—data handling Task 2—mapwork (including GIS application) Task 3—research/essay writing (including GIS application) Two tests Midyear examination	25
	End of year examination: Paper 1 (theory) = 225 Paper 2 (mapwork & GIS) = 75	75
12	Task 1—data handling Task 2—mapwork Task 3—research/essay writing Two tests Midyear examination Trial examination	25
	End of year examination: Paper 1 (theory) = 225 Paper 2 (mapwork & GIS) = 75	75

Table 8.3 GIS weightings of assessment in the FET phase (Department of Basic Education 2011)

Grades	Examination	Paper 1 Total	Paper 2 Total
10	Mid-year End of year	140 225	60 (GIS is 10 marks) 75 (GIS is 15 marks)
11	Mid-year End of year	225 225	75 (GIS is 15 marks) 75 (GIS is 15 marks)
12	Mid-year End of year	225 225	75 (GIS is 15 marks) 75 (GIS is 15 marks)

8.5 The Concepts of Continuity, Sequencing, and Progression

One perspective on quality in geography education has been advanced by Weedon and Butt (2009) who argue that the triad of continuity, sequencing, and progression are key to developing geographical thought amongst learners. As this chapter uses continuity, sequencing, and progression to unpack GIS assessment, they will now be elaborated.

It is argued that there are numerous dimensions to the concept of continuity, but it is largely perceived in education as a dimension of curriculum design; a valuable construct for curriculum developers (Ornstein and Hunkins 2009; Nicolescu and Petrescu 2015). Owen and Ryan (2001) assert that continuity is evident in a long-term plan where there is consistency in content themes (local, global), approaches such as inquiry learning, and in skills acquisition. Gorwood (2018) links continuity to sequencing, maintaining that continuity encompasses one valuable experience following another. In this way, significant learning develops over time related to critical concepts and inquiry/skills in geography. He further links continuity to progression, explaining that continuity is embedded in the linear architecture of subjects where new concepts and skills are firmly grounded in prior knowledge on a topic. Literature, however, on continuity in assessment is seemingly absent.

It has been asserted that there is a gap in academic literature on the concept of sequencing (O'Neill et al. 2014) despite it being one of the dimensions of curriculum design used by curriculum developers (Ornstein and Hunkins 2009). Sequencing in educational literature refers to the order in which the content, inquiry, and skills are presented. O'Neill et al. (2014) explain the 'many different forms of sequencing: simple to complex, easy to difficult, prerequisite learning, whole to parts and vice versa, chronological, developmental, and known to unknown' (p. 270). Although much has been written about programme sequencing and threshold concepts (Land et al. 2008), there appears to be scant literature on assessment sequencing in education. Nevertheless, there is mention of the importance of sequencing, that is, attention to the planning of assessments across the years, that is long-term planning (O'Neill 2010).

The concepts of sequencing and progression are also closely connected; for example, progression is explained as being about the provision of experiences which are sequential in developing learners hierarchically in terms of 'content, context and interactions' (Wood and Bennett 1999, p. 9). Other scholars have also explained that progression involves the thoughtful sequencing of content, inquiry, and skills for learners such that these develop over time (Chambers and Donert 1996; Butt 2002; Bennetts 2005). The literature illuminates some common design aspects of progression in a curriculum. These include increasing the breadth of geographical knowledge, increasing depth of geographical understanding, and increase/wider range in the scale of places studied; an increase in the complexity studied; use of more abstract ideas; an increasing awareness and understanding of social, political, and environmental issues and attitudes; and the development and use of geographical skills with more precision in intellectual and practical tasks (Owen and Ryan 2001; Rawling 2001). Hence, for social constructivists, progression can be seen 'in the nature of the task' (Wood and Bennett 1999, p. 9).

In respect to continuity, sequencing, and progression advancing quality in education, Weedon and Butt (2009) warn that it is possible to have a badly designed geography curriculum with some continuity in terms of geographical content etcetera, but with poor attention afforded to its progression. Good sequencing is evident when there is an attempt to establish the assemblage of the content with aligned activities in a manner which scaffolds the learning process (Bennetts 2005; Weedon and Butt 2009).

8.6 Continuity, Sequencing, and Progression in the GIS Assessments in the CAPS

In analyzing the assessments in GIS in CAPS using continuity, sequencing, and progression, the following elements and trends were found.

8.6.1 Continuity

Continuity is evident in the GIS assessments in the CAPS. There is a total of six assessments for Grades 10 and 11 and seven assessments for Grade 12. There is continuity in that the assessments are the same from Grades 10 to 12 with one exception. There is a seventh assessment in Grade 12 which is a trial examination. GIS is, therefore, tested in all three grades (10–12) within the main summative assessments in the year; that is, the mid-year and final examinations as well as the trial for Grade 12. A perusal of the Grade 11 and Grade 12 Paper 2 (mapwork & GIS) indicates that these papers mirror each other in format and mark allocation. This clearly signals that the Grade 11 Geography Paper 2 in the end of year examination is a preparation for the secondary school exit examination paper in Grade 12, again indicating continuity. However, as shown in Table 8.2, continuity is absent for Tasks 2 (mapwork) and 3 (research/essay writing). In Grades 10 and 11, these include an application of GIS which does not feature in Grade 12. This shows that the research/essay writing is 'learning about GIS' rather than 'learning through GIS'. This also has implications for the sequencing and progression of GIS assessments.

Whilst continuity is a valuable concept towards improving the quality of education in South Africa, it is critical to highlight that the quality of assessments in general and GIS more specifically are impacted upon by existing educational realities such as inequalities in resourcing prevalent in public schools in rural and township areas. Continuity in GIS is compromised as not all learners have access to the required technology to achieve GIS competence. Due to these resourcing issues, disadvantaged schools often select the GIS essay option ('learning about GIS') whilst advantaged schools are propelled towards the GIS research option ('learning through GIS'). Thus, maintaining the same level of quality for all learners in GIS teaching and assessment is clearly lacking.

8.6.2 *Sequencing and Progression*

As already mentioned, in the Grade 12 examinations (Paper 2, mapwork & GIS), all questions are 'about GIS'; there is no assessment 'through GIS'. In other words, while the curriculum document calls for the application of a variety of GIS techniques, often these skills are taught through a paper-based approach because of limited access to correct technology. This not only has implications for continuity but also suggests limited sequencing and progression in terms of GIS application. This is contrary to the general aims of the CAPS which requires learners to identify and solve problems, to be creative and critical, and use technology effectively. Additionally, there is only a very minor increase (5 marks) in marks allocated between Grades 10 and 11 with no further changes in marks allocated for Grade 12 (see Table 8.3). This may indicate an issue with progression in terms of the complexity and use of geographical skills such as GIS (Owen and Ryan 2001). Sequencing and progression are also interrupted with the addition of a totally new assessment in Grade 12 in the form of the trial exam. Furthermore, with little differences in the descriptors used on criteria sheets for Tasks 1–3 across the grade levels, more issues with sequencing and progression are evident. Despite some evidence of continuity, sequencing, and progression as markers of quality education in the CAPS, in order to address the remaining inequalities, I put forward three recommendations for the future: (1) parity in resources through partnerships; (2) professional learning; and (3) integration of assessment *for* learning and *of* learning in the context of GIS questions.

8.6.2.1 Parity in Resources Through Partnerships

While scholars such as Akinyemi (2015) and Le Grange and Ontong (2018) have applauded the inclusion of GIS in school geography, there have been many studies documenting the challenges associated with its implementation. Schools lack infrastructure, such as access to electricity (and hence wifi) and funding to purchase hardware such as computers and GIS software (Van der Westhuizen and Fleischmann 2018; Tarisayi 2018; Malobola 2021; Hlatywayo 2021; Zondi and Tarisayi 2020). This results in the absence of practical application of GIS in numerous South African public schools which are predominantly Black (Fleischmann and Van der Westhuizen 2016; Ahiaku and Mncube 2018; Malatji and Singh 2018; Malobola 2021; Hlatywayo 2021). For example, Van der Westhuizen and Fleischmann (2018, p. 395) summarized this situation by claiming that 'the disparity between the haves and the have-nots in South African education could not be plainer than in the utilization of GIS practice within the class'.

Achieving quality in education, and specifically in the use of GIS in geography, cannot be realized whilst inequalities amongst public schools persist. It is the government's responsibility to provide funding for basic physical resources for GIS teaching, learning, and assessment. However, to date, governments appear unable to equip public schools with the funding necessary for resourcing. To counter this, I argue that private–public partnerships are necessary to assist governments to lift disadvantaged schools out of the current GIS resources imbalance. Schools have governing bodies which are responsible for fundraising, and they should make attempts to garner appropriate local business investment. Without proper financial partnerships going forward, the most disadvantaged schools with high populations of Black learners will remain, as Spaull (2013) has argued, in a bifurcated education system still rooted in apartheid (see also Hay 2018).

8.6.2.2 Professional Learning

Not only do many public schools remain trapped in a discourse of inadequate physical resources, various studies have indicated human resources are lacking; many teachers lack professional learning in GIS implementation or have not been adequately trained in the first place. For example, Malobola's (2021) study in one district of a rural province of South Africa revealed that newly qualified teachers lacked confidence and capacity to implement GIS in their classrooms due to limited exposure to GIS training at university. Therefore, it is necessary to encourage local higher education institutions to implement curriculum reform in their pre-service and in-service teacher training programmes in geography to include GIS teaching, learning, and assessments. Only when schools are properly resourced in terms of infrastructure, technology, and trained staff will continuity, sequencing, and progression for learning GIS be equal for all.

8.6.2.3 Integration of Assessment *for* Learning and *of* Learning in the Context of GIS Questions

While the CAPS document highlights assessment *for* learning as necessary in the FET phase in geography, presently assessment *of* learning is still favoured in the form of high stakes examinations with high mark allocations. This is problematic when cohorts of learners, particularly those in public schools, do not attempt to answer any questions in the GIS section (Fleming 2016). For example, the Eastern Cape Province geography examination report (Eastern Cape Department of Education—Geography specialist inspector 2021) (a rural province of South Africa with a majority of Black African learners) stated that 'in spite of … upward movement in learner performance, there are questions in which learners struggled tremendously. Learners … lack the ability to apply these concepts to deal with real life situations.

Questions 4.1.2 (relevance of vector using vector data) and Q 4.2.2 (application of scale to a geographical problem) are classic examples'. The report noted that 'most learners … lacked basic knowledge/content concerning GIS', a finding that resonates in other studies and contexts (see Zondi and Tarisayi 2020; Hlatywayo 2021).

If assessment *for* learning received more weighting in contributing to the learners' final mark, not only would continuity be enhanced but there may be improved outcomes for students' performances. There is thus a need for more formative assessments (assessment *for* learning) to be included in the assessment programme across the FET for progression in GIS learning. Additionally, a related recommendation may be the need to integrate GIS from the GET phase so there is continuity into the FET.

8.7 Conclusion

The national government in South Africa has taken few small steps in transforming the apartheid school curriculum to one which is more democratic and aligned to global and local needs by including themes such as GIS. However, the position of GIS in geography education can be described as contested. This chapter has highlighted that while there is some evidence of continuity, sequencing, and progression related to GIS assessment in geography, there is a lack of commitment to funding resources and professional learning activities required for the equitable use and assessment of GIS in all public schools. Paper-based assessment in GIS, in particular, appears to be a politically myopic endeavour towards acknowledging the needs of the fourth industrial revolution which requires practical GIS training. The inclusion of GIS in the FET curriculum in part exacerbates the socio-economic inequalities prevalent in disadvantaged public schools and the disparity in resources needed (tools and teacher training) to ensure quality education for all learners including learners who are poor, Black, and bounded by the vestiges of an apartheid history.

References

Ahiaku PKA, Mncube DW (2018) Geography educators' perceptions of learner performance in Grade 12 geography in public schools. Alternation Special Edition 21:68–90

Akinyemi FO (2015) An assessment of GIS use for teaching in Rwandan secondary schools. Geogr Teach 12(1):27–40

Beets P, Le Grange L (2008) Has geography curriculum reform in post-apartheid South Africa strengthened continuity and progression? S Afr Geogr J 90(2):68–79

Bennetts T (2005) The links between understanding, progression and assessment in the secondary geography curriculum. Geogr 90(2):152–170

Butt G (2002) Reflective teaching of geography 11–18: continuum studies in reflective practice and theory. Continuum, London

Chambers B, Donert K (1996) Teaching geography at Key Stage 2. Chris Kington Publishing, Cambridge

Chisholm L, Volmink J, Ndhlovu T, Potenza E, Mahomed H, Muller, J, Lubisi C, Vinjevold P, Ngozi L, Malan B, Mphahlele L (2000) A South African curriculum for the twenty first century: report of the Review Committee on Curriculum 2005. Department of National Education, Pretoria

Christie P (2008) Opening the doors of learning: changing schools in South Africa. Heinemann, Johannesburg

Department of Basic Education (2011) Curriculum and Assessment Policy Statement, Grades 10–12, Geography. Republic of South Africa. https://www.education.gov.za/Portals/0/CD/National%20Curriculum%20Statements%20and%20Vocational/CAPS%20FET%20_%20GEOGRAPHY%20_%20GR%2010-12%20_%20WEB_C9A9.pdf?ver=2015-01-27-154144-177

Department of Education (1995) White paper on education and training in South Africa: first steps to develop a new system (White Paper One). Government Gazette, Pretoria

Department of Education (2005) The national protocol on assessment for schools in the general and further education and training band (Grades R–12). Republic of South Africa. https://www.gov.za/sites/default/files/gcis_document/201409/doefetnationalprotocolassgrader12.pdf. Accessed 10 June 2021

Department of Education (2012) Curriculum and Assessment Policy Statements (CAPS). Republic of South Africa

Eastern Cape Department of Education—Geography specialist inspector (2021, July 25) Geography Report [personal communication with Geography Education District Head]

Eiselen W (1953) Report of the Commission on Native Education 1949–1951. Government Printer, Pretoria

Fleisch B (2002) State formation and the origins of Bantu Education. In: Kallaway P (ed) The history of education under apartheid 1948–1994: the doors of learning and culture shall be opened. Pearson, Cape Town, pp 39–52

Fleischmann EM, Van der Westhuizen CP (2016) A GIS integrational framework for poorly resourced schools. J Geogr Educ S Afr 1(2):25–33

Fleming B (2016) GIS interventions at secondary level education in South Africa—some recent successes and shortfalls. J Geogr Educ S Afr 1(2):9–24

Gorwood BT (2018) School transfer and curriculum continuity. Routledge, London

Hay AA (2018) Quality education: a school perspective. Masters in Education, University of KwaZulu-Natal, South Africa. https://researchspace.ukzn.ac.za/bitstream/handle/10413/16982/Abdool%20Hay_Ayesha_2019.pdf?sequence=1&isAllowed=y

Hlatywayo J (2021) The pedagogical approaches to teaching GIS. PhD thesis (submitted for examination), University of KwaZulu-Natal, South Africa

Jansen JD (1999) The school curriculum since apartheid: intersections of politics and policy in the South African transition. J Curric Stud 31(1):57–67

Jansen JD (2002) Political symbolism as policy craft: explaining non-reform in South African education after apartheid. J Educ Pol 17(2):199–215

Jansen J, Christie P (eds) (1999) Changing curriculum: studies on outcomes-based education in South Africa. Juta & Co, Kenwyn

Land R, Meyer JHF, Smith J (2008) Threshold concepts within the disciplines. Sense Publications, Rotterdam

Le Grange L, Beets P (2005) Geography education in South Africa after a decade of democracy. Geogr 90(3):267–277

Le Grange L, Ontong K (2018) Towards an integrated geography school curriculum. Alternation Special Edition 21:12–36

Lentsa N (2019) A theory-based evaluation of South Africa's education policy. Masters in Arts in Political Science, Stellenbosch University, South Africa

Malatji KS, Singh RJ (2018) Implications of the articulation gap between Geography learners in secondary schools and university. Alternation Special Edition 21:91–108

Malobola ZM (2021) The professional experiences of newly qualified geography teachers. University of KwaZulu-Natal, South Africa, PhD

Manik S (2016) Tracing geography education's footprints in South Africa. In: Visser G, Donaldson R, Seethal C (eds) The origin and growth of geography as a discipline at South African Universities. African Sun Media, Cape Town, pp 445–465
Motala S (2011). Educational access in South Africa. J Educ Stud (Special Issue Social Justice). https://journals.co.za/doi/https://journals.co.za/doi/10.10520/EJC160137
Nel E, Binns T (1999) Changing the geography of apartheid education in South Africa. Geogr 84(2):119–128
Nicolescu BN, Petrescu TC (2015) On the continuity mathematics curriculum between primary and secondary school. Procedia Soc Behav Sci 180:871–877
O'Neill G (2010) Initiating curriculum revision: exploring the practices of educational developers. Int J Acad Dev 15:61–71
O'Neill G, Donnelly R, Fitzmaurice M (2014) Supporting programme teams to develop sequencing in higher education curricula. Int J Acad Dev 19(4):268–280
Ornstein AC, Hunkins FP (2009) Curriculum foundations, principles and issues, 5th edn. Allyn and Bacon, Boston
Owen D, Ryan A (2001) Teaching geography 3–11. Continuum, London
Rawling E (2001) Politics and practicalities of curriculum change 1991–2000: issues arising from a study of school geography in England. Br J Educ Stud 29(2):137–158
Sayed Y (2002) Educational policy in South Africa: from opposition to governing and implementation. Int J Educ Dev 22(1):29–33
Schwab K (2017) The fourth industrial revolution. Crown Business, New York
Spaull N (2013) South Africa's education crisis: the quality of education in South Africa 1994–2011. CDE, Johannesburg
Tarisayi KS (2018) Lessons for GIS implementation in Zimbabwe. Alternation Special Edition 21:185–202
Union of South Africa (1953) Bantu Education Act, Act No. 47 of 1953. http://psimg.jstor.org/fsi/img/pdf/t0/10.5555/al.sff.document.leg19531009.028.020.047_final.pdf
Van der Westhuizen CP, Fleischmann EM (2018) GIS, Google Earth, Google Maps and GPS as resources and their application in Geography Education. In: van Eeden E, Warnich P (eds) Teaching and learning history and geography in the South African classroom. Van Schaik, Hatfield, pp 389–432
Van Harmelen U (1999) Where has all the geography gone? A social constructivist perspective of Curriculum 2005. S Afr Geogr J 81(2):80–85
Weedon P, Butt G (2009) Assessing progress in your Key Stage 3 geography curriculum. Geographical Association, London
Wilmot D, Dube C (2015) School geography in South Africa after two decades of democracy: teachers' experiences of curriculum change. Geogr 100(2):94–101
Wood E, Bennett N (1999) Progression and continuity in early childhood education: tensions and contradictions. Int J Early Years Educ 7(1):5–16. https://doi.org/10.1080/0966976990070101
Zondi T, Tarisayi K (2020) A learner perspective on the implementation of Geographic Information Systems in selected schools in KwaZulu-Natal. J Transdisciplinary Res S Afr. https://td-sa.net/index.php/td/article/view/752/1258

Sadhana Manik is an Associate Professor in geography education in the School of Education at the University of KwaZulu-Natal. She teaches at both postgraduate and undergraduate levels and leads two international staff, school, and student exchange programmes with HVL of Norway. She received a 'teaching excellence' award in 2018, and in 2019 she was ranked in the top 30 list of researchers who achieved 'excellence in research' within the College of Humanities at the University of KwaZulu-Natal. Her research interests include teaching, learning, and assessment in Geography education, access and success of students in higher education, and international teacher migration.

Part V
Future Directions for Assessment in Geography Education

Chapter 9
Online Adaptive Assessment of Map Skills

Martin Hanus and Lenka Havelková

Abstract The dispersion of new technologies into education opens up new perspectives for the assessment of students' performance, not only in their knowledge but also in their skills. One of these perspectives is the use of systems that enable adaptive assessment, i.e., personalized (mostly online) assessment according to the actual position of each student in the knowledge or skills hierarchy. The general presumption of this approach is that each student solves tasks according to his/her previous success—if the student fails, then s/he is presented with lower-level tasks; if s/he succeeds, s/he proceeds to more complicated tasks. Students' progress is supported by the hints available and in particular by the structure of tasks, as easier tasks support the solution of the more complicated ones according to the theory of pedagogical scaffolding. The chapter first discusses a general framework for the use of adaptive assessment and then presents a practical example of the use of the framework to assess the level of map skills—the online adaptive application Mapwork.education. Following a general description of the application, various possibilities of its use in geography lessons are outlined.

Keywords Geography education · Adaptive learning · Adaptive testing · Online assessment · Map skills

9.1 Introduction

We live in a time of a dynamically developing society, in economic, social, cultural, and, above all, technological areas. Global processes are constantly accelerating

M. Hanus (✉) · L. Havelková
Faculty of Science, Charles University, Prague, Czechia
e-mail: martin.hanus@natur.cuni.cz

L. Havelková
e-mail: l.havelkova@natur.cuni.cz

T. Bourke et al. (eds.), *Assessment in Geographical Education: An International Perspective*, Key Challenges in Geography,
https://doi.org/10.1007/978-3-030-95139-9_9

this development, which brings new challenges for education systems around the world. If (geographical) education is to 'produce' active and responsible (geographically thinking) citizens, it must reflect these changes in society. It means adapting the educational goals towards society's needs, employing technologies to increase the efficacy of learning, and developing teaching and assessment strategies appropriate to the changing frameworks of education. One of the progressive approaches employs computer-based technologies to address students' individual needs and develop adaptive and adaptable strategies of learning and assessment.

Vie et al. (2017) stress the difference between adaptivity, i.e., the ability of teachers or educational systems to modify course materials using different parameters and a set of pre-defined rules, and adaptability, i.e., the possibility for learners to personalize the course materials by themselves. Moreover, they call for a better balance between giving learners what they need to learn (i.e., adaptivity) and giving them what they want to learn (i.e., adaptability). Both aspects of learning and assessment are traditionally overlooked in (geography) lessons (Lane and Bourke 2019).

Students learn in different ways and at different levels. Some students may be strong in one academic field but weak in others. This is experienced by teachers at all levels of education (from primary to tertiary), especially in the first years of study after students are promoted from a lower level of education (or after a transfer from another school). Students may have acquired knowledge from diverse backgrounds, and some may be missing several prerequisites, whereas others could afford to skip certain parts of the lesson. Therefore, it would be useful to recognize individual students' needs and preferences, adapt the educational content accordingly, and minimize information overload (Vie et al. 2017). Wongwatkit et al. (2017), following Raman and Nedungadi (2010), have, therefore, emphasized the necessity of providing personalized learning guidance or support by considering individual students' achievement levels, needs, and learning styles to improve their learning performances.

There is now a more widely shared view that assessment should fulfil a plurality of purposes (Challis 2005). Nightingale et al. (1996) pointed to three pressures for change: to develop (and consequently to assess) a much broader range of student abilities; the desire to harness the full power of assessment and feedback in support of student learning; and to involve students in the process of assessment to develop independent judgement and the ability to evaluate one's own performance. Moreover, the assessment process should help students be responsible for their own learning (Wang 2010).

Nonetheless, it is difficult to closely observe a particular student's progress, especially in large classes. Guiding students in their learning, grading them on their knowledge/skills level, and providing individual feedback is challenging and time consuming for teachers (Figueira-Sampaio et al. 2009; Martín and Rubio 2009; Chatzopoulou and Economides 2010; Raman and Nedungadi 2010). Providing personalized learning and feedback is even more challenging in face-to-face teaching. However, with advances in computing technology, integrating computer-assisted assessment and feedback into lessons allows students to individually learn (geography), determine what their learning flaws are, and improve their conceptual understanding and metacognition (Hsiao et al. 2018).

In the past decade, researchers have attempted to develop testing and diagnostic systems to detect students' learning levels and causes of learning failures and to suggest remedial learning paths for individual students (for an overview see, e.g., Wongwatkit et al. 2017). Many systems have recently been developed to help students learn in various disciplines, including mathematics, natural science, geography, and computer science (Hwang et al. 2013; Wongwatkit et al. 2017). One of the first attempts at adaptive learning and assessment in geography (specifically, in the field of map skills) was carried out by van Dijk et al. (1994). However, the use of adaptive systems in geography education is still rather rare (Lane and Bourke 2019).

Considering this lack of attention to adaptive assessment in geographical education and the evident benefits that adaptive assessment brings, this chapter discusses a general framework for the use of computer-based adaptive online assessment. Subsequently, it presents a practical example of adaptive assessment of map skills—the online adaptive system Mapwork.education. Finally, some recommendations for the system's use in geography lessons that are (in addition) transferable to other adaptive systems are presented.

9.2 Adaptive Learning

The benefits of adaptive learning and assessment are closely linked to benefits of using computer-based technologies for educational purposes. As early as the end of the twentieth century, Brown and Knight (1994) called for the exploitation of computer networks' power to transfer, upload, and download data automatically to benefit student learning by receiving specific, timely, and encouraging feedback. Since then, research has attempted to propose learning systems that have the advantages of monitoring and assisting students as they engage in learning activities (Chu et al. 2011). These systems especially aim to adjust learning content according to individual students' cognitive levels to meet their needs—i.e., adaptive learning (Chen 2011; Wang 2014; Ting and Kuo 2016).

More specifically, the learning is adapted according to learners' prior knowledge. This is believed to be one of the most important factors affecting learning effectiveness as it impacts learner achievement (Dochy 1994). Dochy et al. (2002) and Hailikari et al. (2007) argued that prior knowledge interacts with different information processing phases. Learners lacking appropriate prior knowledge will have trouble learning new information and constructing new understandings (Hailikari et al. 2007; Wang 2010). Wu and Chang (2020) confirmed the increased effectiveness of adaptive learning by students exhibiting better learning performance and by a higher ratio of misconception revisions. Moreover, students learning adaptively spent significantly less time on learning (in contrast to a non-adaptive group).

To overcome the difficulty of adaptive learning implementation, it is necessary to incorporate a formative assessment method that monitors and provides feedback during the learning process, regardless of the users' specific conceptual learning problems, learning paths, or learning styles. These systems would then be more useful for improving students' learning achievement. Therefore, learning systems have to provide not only the content but also appropriate feedback during the learning process (Nacheva-Skopalik and Green 2016; Wongwatkit et al. 2017).

9.3 Adaptive Assessment

Although adaptive assessment has existed in the educational environment since the time of oral examinations, its full potential can be currently exploited through advanced technologies (Challis 2005), offering speed, availability, consistency, and objectivity of the assessment (Ala-Mutka 2005; Chatzopoulou and Economides 2010). Conventional computer-assisted assessment facilitates a quick calculation of student assessment scores, but the results do not reveal students' cognitive levels and they provide little information for teachers to improve the learning process (Lau and Yuen 2010; Chen 2011). In contrast, learning analytics for adaptive assessment have specific and well-defined objectives: they must improve the learning process's efficiency and effectiveness and tell learners what to do next by adaptively organizing instructional activities (Chatti et al. 2012). The elementary framework of the adaptive assessment system is explained by Challis (2005, p. 521):

> The distinctive element of adaptive testing is that, as the student responds to test items, the items change to reflect the performance on the preceding item(s). This means that any adaptive test is designed based on the requirement that the user's level is constantly established. Commencing at a moderate level, as the user answers questions, items become more difficult/complex or easier, depending on the demonstrated level of answers. The score is not derived from the number of correct answers but, rather, the level of difficulty of the questions answered correctly. Such tests can be designed to be terminated when a student has reached a required level of proficiency/competency, or that performance has reached the user's highest level.

Therefore, the goal of the adaptive assessment system is to appropriately filter the questions to be asked. Instead of asking everyone the same questions, the adaptive systems select the next question based on the previous answers, thus allowing adaptivity at each step. The adaptive tests' design relies on two criteria: a termination criterion (e.g., a number of questions to ask) and a next item criterion (van der Linden and Glas 2000; Vie et al. 2017). Vie et al. (2017) stress that assessment of skills must be performed under uncertainty: a learner may slip, i.e., accidentally or carelessly fail an item that they could have solved, or they may guess, i.e., correctly answer an item by chance. This is why adaptive assessment cannot simply perform a binary search over the learner's ability, i.e., asking a more difficult question if they succeed once and an easier question if they fail once.

Students' success or failure on a learning task should be assessed based on whether they master (or fail to master) key knowledge components—prerequisites. These knowledge components allow a transfer of evidence from one item to another (Vie et al. 2017). For the adaptive tests, specification of which knowledge components are required to be solved in each test item is needed according to the Attribute

Hierarchy Model (Leighton et al. 2004). An adaptive assessment can then uncover the students' level and suggest the next knowledge components to learn to help them progress through a learning path (Vie et al. 2017).

Considering the findings of previous studies, the benefits of adaptive assessment can be highlighted.

Benefits for learning:

- Adaptivity, adaptability, and feedback, as learners can observe their personal learning progress and decide how further to direct their learning process, while teachers can individually support learners and formulate judgements about the quality and effectiveness of the provided instruction (Gouli et al. 2002; Challis 2005).
- Efficiency, as the attained level can be determined with fewer questions, because the generated questions are tailored to learners' level of proficiency, and so they do not have to attempt numerous questions that are either too easy or too difficult for them (Gouli et al. 2002; Lynch and Howlin 2014; Chen et al. 2015).
- Accuracy, as a more accurate estimation of the learners' proficiency level can be achieved compared with conventional methods (Challis 2005).
- Cognitive variability, as it is possible to sufficiently assess a wider range of learning outcomes such as analysis, synthesis, evaluation, and application because of the reduction of inadequate questions (Brown and Knight 1994).
- Appropriateness, as learners specify the level at which they wish to work in the lesson (Challis 2005).
- Attractivity and motivation, as adaptive assessment enhances learners' performance, attracts their attention, and increases their learning motivation (Foster et al. 2016; de-Marcos et al. 2016).
- Attentiveness, as shorter tests are better for learners, who may be frustrated or bored if they need to give too many answers (Lynch and Howlin 2014; Chen et al. 2015).
- Privacy and security, as online adaptive assessment can be conducted in a private space with freedom to explore areas of perceived weakness and to make mistakes (Challis 2005).
- Time appropriateness, as, in most cases, systems of adaptive assessment can be accessed whenever convenient for the learner (Challis 2005).
- Reliability, as adaptive assessment lessens the likelihood of learners cheating (Challis 2005).
- Self-efficacy, as learners' satisfaction is higher when working at the appropriate level (Chen 2011; Wongwatkit et al. 2017).

Benefits for teaching:

- Courses can be designed to facilitate immediate provision of feedback to learners on how successfully they understand and apply key concepts (Challis 2005).
- Where gaps and weaknesses are exposed, remediation can be provided through embedded feedback and assistance, which can be prepared once and then accessed as frequently as needed (Challis 2005).

- Adaptive assessment prior to instruction can result in a reduced need for repetitive drills before learners are confident that they have the requisite level of competency (Challis 2005).
- Teaching based on prior knowledge helps learners to move forward faster than in traditional teaching (Raman and Nedungadi 2010).

As with other assessment methods, adaptive assessment must distinguish whether it serves as a summative assessment (to certify the knowledge/skills level of students) or is conducted for formative purposes to assist learning progression. Traditionally, adaptive assessment models were mostly summative: they measured or ranked examinees effectively without providing any other feedback (e.g., models derived from the item response theory; Hambleton and Swaminathan 1985). Currently, the greatest promise is in the formative assessment area (Challis 2005), providing useful feedback to assist teaching and learning (Vie et al. 2017).

9.4 Map Skills in Geography Education

To illustrate how online adaptive assessment can be efficiently used in geography education, the Mapwork.education system (Centre for Geographical and Environmental Education 2020) is presented in the following text (see Sect. 9.5). Mapwork.education was primarily developed for both formative and summative assessment of students' map skills level at lower and upper secondary schools (aged 11–19).

Map skills, i.e., skills needed for efficient use and design of maps, are a fundamental part of geography education in all levels (Catling 1980, 2020; Young 1994; Hanus and Havelková 2019; Briwa and Wetherholt 2020). Nowadays, more than ever, the need to be able to critically evaluate and interpret visualized information is apparent (Dörk et al. 2013; Kosminsky et al. 2019). Only with fully mastered map skills can students recognize the power that maps have and appropriately adapt their conclusions, predictions, and decisions (Kitchin and Dodge 2007; Cinnamon 2017; Briwa and Wetherholt 2020).

The skills for effective and full use of already created maps are developed gradually as they differ in their complexity and therefore in their difficulty (van Dijk et al. 1994; Havelková and Hanus 2018). Many experts distinguish three main types (levels) of map use skills based on which individual map skills are classified: map reading, map analysis, and map interpretation (e.g., Wiegand 2006; Kimerling et al. 2016; Hanus and Havelková 2019; see Fig. 9.1). Distinguishing the main types of map skills, structuring them into a hierarchical cognitive model, and specifying their knowledge/operational components is fundamental for the development of effective adaptive assessment systems.

Map reading skills, referring to simple identification of individual objects and locations and decoding of map symbology, are a prerequisite for acquiring map

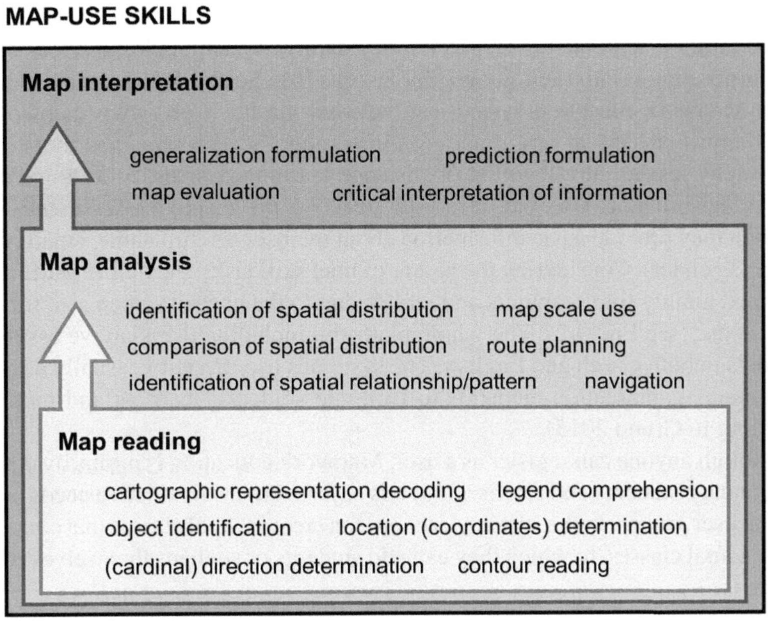

Fig. 9.1 Basic map use skills and their categorization

analysis and map interpretation skills. Likewise, students need to master map analysis skills, which mainly relate to spatial distribution of objects/phenomena and navigation with a map, to be able to interpret maps. When students develop map interpretation skills, they are able to formulate generalizations and predictions based on maps, evaluate them, and critically interpret displayed information (Fig. 9.1).

In geography education lessons, it is necessary to incorporate map use together with map design focused on various topics. And, in particular, to answer geographic questions and solve geographic problems. Map skills, therefore, should not stand alone as the main subject and aim of a few geography lessons. For a detailed explanation of effective map work in geography lessons, see Hanus and Havelková (2019).

9.5 Mapwork.education—A System for Online Adaptive Assessment

Mapwork.education is a web-based, fully automated adaptive assessment system. It is part of an extensive platform aiming to help Czech geography teachers with the systematic and efficient development of students' cartographic knowledge and map skills. For instance, an integral part of the platform is the teachers' handbook titled *Map Work in Lessons*, approved (as a certified methodology) by the Czech

National Pedagogical Institute. The handbook, among other things, helps teachers become familiar with the developed online adaptive system and introduces options for incorporating it into their geography lessons (see Sect. 9.5.2 as well).

The Mapwork.education system can be used via the Web (www.mapwork.education) anywhere and at any time, e.g., in a school's computer laboratory during a geography lesson, in e-learning or distance learning courses, or at home during students' leisure time. Users must register to access the adaptive assessment system, for which they provide basic information about themselves (full name, email, year of birth, and gender). Considering the nature of map skill tasks, the online platform was developed mainly for computers and tablets due to the needed screen size for effective map use (see Fig. 9.2). The whole platform, including the adaptive assessment system, is in both Czech and English. The system is user friendly, visually attractive, and contains gamification elements to facilitate students' learning and motivation (according to Grund 2015).

Although anyone can register as a user, Mapwork.education is primarily designed for secondary students and teachers and it is adjusted to the distinctive needs of these two core user groups. In the system, teachers can create school classes that correspond to their actual classes, to which they can add students or students themselves can join their class.

Mapwork.education provides immediate feedback to students and their teachers regarding their answers and a detailed performance statistics report. Students therefore know exactly which tasks they solved correctly/incorrectly, and they can view their average success rate in all solved tasks, in each task set, for each map skill type (map reading, analysis, and interpretation) and each map type (see Sect. 9.5.1 for more details). To increase students' motivation to continue solving tasks in the system and, therefore, to improve in the tested skills, students are assigned, based on their success, to a certain rank of a fictional 'cartographic ranking' system (from a rookie to a cartographer) and they know how much they need to improve to be 'promoted'. Moreover, a fictional map that resembles an old pirate treasure map is gradually revealed to students based on the number of correctly solved tasks requiring map analysis and map interpretation, in order to motivate their progression in these higher-order skills.

Teachers see the performance statistics not only for themselves but for all students assigned to their classes. In addition, they can see detailed results for each task in the system, i.e., how many students from the class solved it, and how successful they were on average and individually. They can also display any task assignment, with given possible answers and the highlighted correct answer, and the map(s) used in the task, whenever they want. This enables teachers to use the Mapwork.education adaptive assessment system as a diagnostic tool and to pay further attention to specific map tasks during lessons. For that reason, teachers, as well as other users, can also display the map skill type and specific map skill that each task is focused on.

In two years since its launch, Mapwork.education has gained more than 3200 registered users, of which almost 500 are teachers who have created more than 200 school classes in the system. A total of 150,000 answers to map tasks were recorded from all users in the first two years.

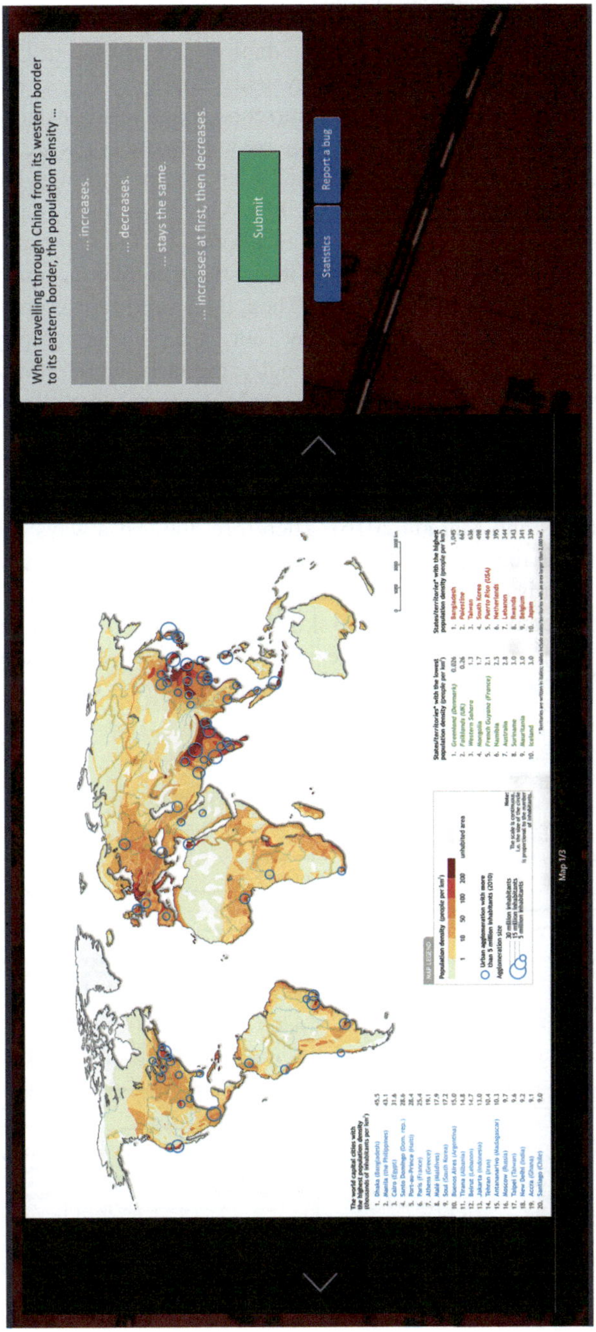

Fig. 9.2 Mapwork.education website interface

9.5.1 Model Structure and Its Content

Currently, Mapwork.education consists of 31 adaptive sets of map tasks, i.e., tests, covering various topics from human and physical geography (e.g., demographic transition, land use, energy sources, and sea currents). Each test consists of eight to 13 map tasks. To enable test adaptivity, all three basic map use skills are represented in the set, i.e., map reading, map analysis, and map interpretation, which defines three difficulty levels of the test tasks (see Fig. 9.3). Altogether, the system task bank consists of 323 tasks. The sets of tasks were developed by the team of geography educators and pre-service teachers according to the theoretical concepts mentioned in Sect. 9.4 (see Hanus and Havelková 2019). Then, their validity was cross-checked by the rest of the team and, subsequently, by four in-service geography teachers (from lower and upper secondary schools). Finally, the tasks were pilot tested on a sample of 500 secondary students to assess the tasks' comprehensibility, reliability, and other characteristics.

Given that Mapwork.education is mainly designed for lower and upper secondary students, it is expected that the majority of users have already mastered map reading skills (see, e.g., Hanus and Marada 2016). Therefore, the test always starts with a task requiring some of the map analysis skills (see Fig. 9.4). If the task is solved incorrectly, the student is given task(s) requiring only map reading skills, i.e., easier tasks (Fig. 9.1). To effectively support map skill development, the given map reading tasks in all tests are interrelated with the incorrectly solved map analysis tasks. For instance, if a map analysis task requires finding the region with the lowest life

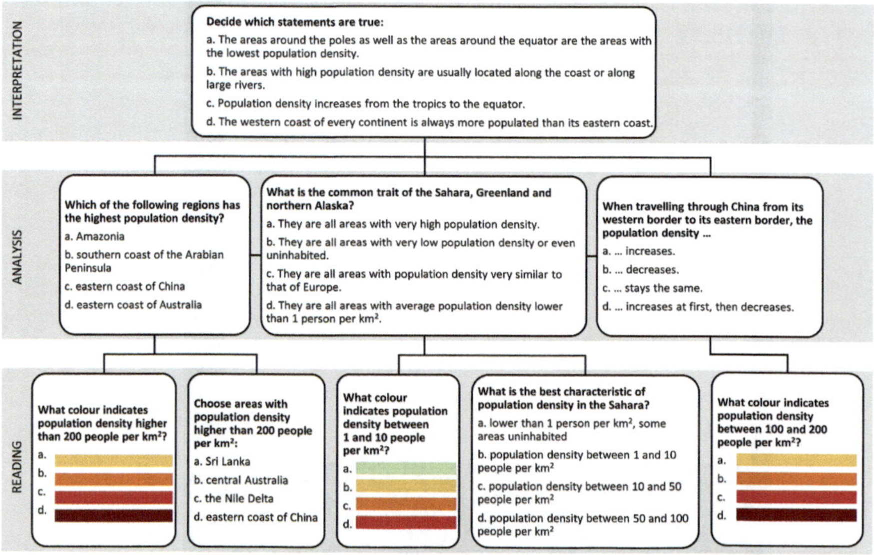

Fig. 9.3 An example model of test task hierarchy (the map needed to solve the tasks is available at http://shorturl.at/gmyL1)

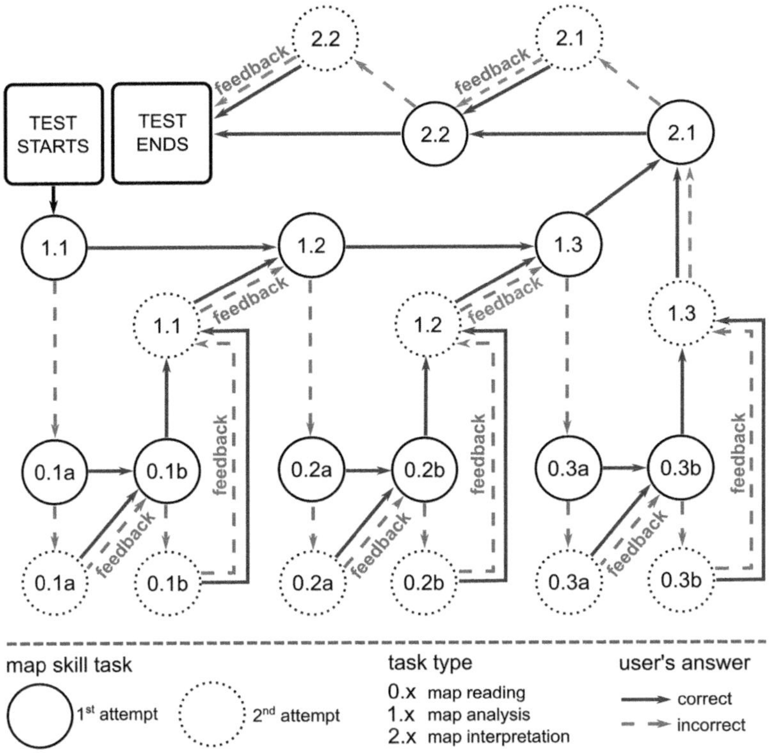

Fig. 9.4 An example model of administered adaptive test (*Note* The number of tasks for each map skill type can slightly differ across tests, but the structure is preserved)

expectancy, the map reading tasks can require the student to identify the colour that represents the lowest values or to state the interval of the life expectancy value in Niger.

If the student fails to solve a map reading task, s/he is given another chance to solve it. After a second incorrect answer of the same map reading task, the correct solution is revealed together with an explanation when necessary. The same system is applied to other map reading tasks related to the first map analysis task, to which the student returns afterwards and has another chance to solve (Fig. 9.4). If s/he is still unable to solve the map analysis task, the correct solution is shown together with an explanation (why the given solution is correct and other possibilities are not). Subsequently, both successful and unsuccessful users of the first map analysis task continue to (an)other map analysis task(s) where the possible set of following steps is identical.

After completing all map analysis tasks in the test, the student is given tasks focused on map interpretation skills (see Fig. 9.4). The student no longer falls to easier tasks after incorrectly answering the map interpretation tasks as s/he already had a chance to prepare for them during the related map analysis tasks (and potentially map reading tasks as well). Hence, the student immediately gets a chance to think about the task again and find out where s/he made a mistake. Similarly to the rest of the tasks, after a second incorrect solution the correct answer together with its explanation is revealed. At the end of the test, the test statistics are shown to the student who can decide if s/he wants to try another test.

The model of the administered adaptive tests was designed in a way that the assessment of students' level of map skills is, at least partially, resistant to students' accidental errors (accidentally or carelessly failed tasks) and guesses (tasks answered correctly by chance). Thanks to its elements of multistage testing, the difficulty level of tasks (type of map skills in this case) does not change after every task but after a set of tasks of the given difficulty level (Vie et al. 2017).

The Mapwork.education system is not only adaptive but also adaptable, and, therefore, it attaches importance to the preferences of the tested students, as their will to learn and progress is crucial (Vie et al. 2017). Not only can students choose where, when, and for how long they will solve a test or if they will take a break during testing, they can also choose the type of map they will use to solve the map tasks. In the system, the most common map types which students use in school and in their personal lives (general reference map, topographic map, choropleth map, proportional symbol map, maps with figural, line, and area symbols) are represented, together with less traditional map types with which students should become familiar as well (cartogram, dot map, dasymetric map, and isopleth map).

The tasks differ, apart from the map type used and the map skill(s) required in the test question type, which can also lead to students' improvement not only in map skills but generally in test taking. The tasks fall into four different types—multiple choice, multiple response, true/false, and gap filling. The multiple choice tasks consist of four possible answers, only one of which is correct. In contrast, the multiple response tasks can have one to four correct answers from four possible answers. To lower the chance that users guess the correct answer, the true/false tasks consist of four statements, and the user must decide which one(s) is/are correct. Gap filling was chosen from open question formats to enable fully automated assessment. These tasks consist of one to four gaps in the text to be filled in. The different task types are distinguished graphically in the system to help students as they take the test.

All tasks in the Mapwork.education system were designed to assess only the level of required map skills (and potentially the prerequisite map skills) and would therefore depend as little as possible on (and be influenced by) students' knowledge. For that reason, not only do users have the map they need to solve the presented tasks, they can also switch, whenever they want, to a general reference or a political map of the world if, for example, they do not know where the country, city, river, etc., mentioned in the task is located.

9.5.2 Use in Education

The Mapwork.education system combines assessment tools with learning instructions (in the form of hints and explanations that can be supplemented by teachers' inputs), emphasizing both its adaptivity and adaptability. This combination results in various possible uses of Mapwork.education in geography education. It is predominantly an educational tool that allows teachers to assess, diagnose, develop, and practice/consolidate students' map skills.

The adaptive assessment system should provide teachers with highly accurate and reliable data on skills that individual students have mastered, instructional needs, students' learning progression, and grouping of students into appropriate instructional programmes. Such data could help teachers make data-driven decisions in the classroom (Raman and Nedungadi 2010). Data from Mapwork.education are, among other things, suitable for teachers' scaffolding of students' building of knowledge, allowing them to shift from occasional scaffolding to planned systematic scaffolding reflecting students' needs and prior knowledge (Wood et al. 1976).

Teachers can obtain all the necessary information from the Mapwork.education system when it is employed as a diagnostic tool, particularly when used in combination with instruction in the model of 'pre-test – instruction – post-test'. According to Sternberg and Grigorenko (2001), there are two formats for its implementation: 'sandwich' and 'cake'. In the 'sandwich' format, students take a pre-test before learning. After completing the pre-test, students receive instruction (the teaching materials are related to the pre-test contents). After the instruction, examinees take a post-test (summative assessment). In the 'cake' format, students receive instruction by answering items one after another. When they answer an item incorrectly, they receive a graded series of hints. These successive hints are designed to help them gradually clarify the answer. The hints constitute instruction, and the instructional activities will not stop until the item is correctly answered.

The choice of format depends fully on the teacher (possibly on students), but both formats emphasize the need for systematic assessment in the whole process of learning. Pre-tests in particular are not commonly used in geography education to diagnose achievement prerequisite or the learning needs of students in geography lessons (Lane and Bourke 2019). This contrasts with the fact that researchers have for many decades emphasized the need to work with students' prior knowledge in courses (e.g., Willcockson 1944; Schlemper et al. 2018).

In addition to the use of Mapwork.education to diagnose pre- and post-instructional levels, it can be employed as formative assessment in the middle of the course. Learners often wish to sample the tasks they will be expected to solve in the final test or practice specific skills. Such assessment should consider the fact that learners' levels may change while they are taking the test, for example, if they check the course material during the test (Vie et al. 2017).

Before using Mapwork.education as a diagnostic tool, teachers should explicitly state the assessment's purpose. This will help them to aim specifically at a deep diagnosis of the use of one specific type of map or a complex assessment of skills related to all available map types.

Additionally, teachers using Mapwork.education can benefit from its learning aspects and, therefore, use it directly for map skill development. Students are constantly learning and developing while solving test tasks (using hints, potentially). Thus, they are improving their map skills. Moreover, as adaptivity allows a completely individual approach, students develop their map skills in their own way. Specifically, the development is adapted to their initial level of skills and to the appropriate pace of learning and consolidating new skills.

Moreover, Mapwork.education can assist learning by following up on teaching and by providing space for practising and consolidating skills developed in the lesson, particularly in the form of homework.

The above-mentioned examples of the use of Mapwork.education in lessons are suitable in almost all forms of teaching—face-to-face, hybrid, and distant. However, teachers have to consider each form's specifics and reflect on them to maintain the quality and efficacy of the learning process.

9.6 Conclusion

Adaptive assessment has considerable potential to become an essential form of assessment in the near or distant future. With respect to this potential and the benefits of adaptive assessment, its wider implementation in (geography) education is limited by several aspects. Such limits result in challenges that must be addressed (by learners, teachers, teacher educators, researchers, curriculum leaders) to fully utilize adaptive assessment potential, aiming at high quality and effective teaching and learning.

In order to implement the adaptive assessment system in the teaching of a wider range of students, it is necessary to have sufficient technical equipment, including particular user skills and the developed hierarchical model of items (according to prerequisites and reflecting key knowledge components). The development of such a system seems to be a cognitively and timely very demanding process that is not easy for teachers to achieve individually for their classes, especially when the assessment is part of a more complex adaptive learning system. In our opinion, the responsibility for maximal support of teachers in implementing adaptive assessment in education should be taken by curriculum leaders together with teacher educators and researchers in geography education. These stakeholders should develop the necessary educational systems, whether as more general software containing adaptive algorithms, which individual teachers could fill with tasks and learning content according to their lesson plans, or as a complex ready-made system (including a technical solution and subject-specific content—i.e., individual test items together with hints and explanations) that fully reflects curricular requirements for a given field and level of education. It would, therefore, be misleading to expect teachers to implement such a complicated and comprehensive assessment without significant external support.

Moreover, the necessary methodological support (how to implement it) should be provided to teachers, especially by the teacher educators in initial teacher education and courses for in-service teachers. In addition to developing the necessary adaptive systems and providing methodological and technical support for teachers, it is necessary to change students' thinking about their own performance assessment.

It is more than obvious that the implementation of adaptive assessment requires changes of the entire educational system, especially in assessment, lesson planning, and teaching. The complexity of these changes is the main limit of adaptive assessment implementation and the reason it is used primarily in distance learning courses at universities. Its implementation in the secondary and primary level of geography education is one of the biggest challenges.

Acknowledgements This study was supported by Charles University under Charles University Research Centre program no. UNCE/HUM/024.

References

Ala-Mutka KM (2005) A survey of automated assessment approaches for programming assignments. Comput Sci Educ 15(2):83–102. https://doi.org/10.1080/08993400500150747
Briwa R, Wetherholt W (2020) Teaching critical cartography in the introductory geography classroom. Geogr Teach 17:162–168. https://doi.org/10.1080/19338341.2020.1828135
Brown S, Knight P (1994) Assessing learners in higher education. Routledge, London
Catling SJ (1980) For the junior and middle school: map use and objectives for map learning. Teach Geogr 6:15–17
Catling SJ (2020) Reflecting on the purpose of mapwork in primary schooling. Int J Cartography 6:270–283. https://doi.org/10.1080/23729333.2020.1770480
Centre for Geographical and Environmental Education (2020) Mapwork.education—adaptive assessment tool. http://mapwork.education/app_v2.php. Accessed 22 Jan 2021
Challis D (2005) Committing to quality learning through adaptive online assessment. Assess Eval High Educ 30(5):519–527. https://doi.org/10.1080/02602930500187030
Chatti MA, Dyckhoff AL, Schroeder U, Thüs H (2012) A reference model for learning analytics. Int J Technol Enhanc Learn 4:318–331. https://doi.org/10.1504/IJTEL.2012.051815
Chatzopoulou DI, Economides AA (2010) Adaptive assessment of student's knowledge in programming courses. J Comput Assist Learn 26:258–269. https://doi.org/10.1111/j.1365-2729.2010.00363.x
Chen L-H (2011) Enhancement of student learning performance using personalized diagnosis and remedial learning system. Comput Educ 56:289–299. https://doi.org/10.1016/j.compedu.2010.07.015
Chen S, Choi A, Darwiche A (2015) Computer adaptive testing using the same-decision probability. In: Proceedings of the Twelfth UAI Conference on Bayesian Modeling Applications Workshop, vol 1565. CEUR-WS.org, Aachen, pp 34–43
Chu K-K, Lee C-I, Tsai R-S (2011) Ontology technology to assist learners' navigation in the concept map learning system. Expet Syst Appl 38:11293–11299. https://doi.org/10.1016/j.eswa.2011.02.178
Cinnamon SA (2017) Map as weapon. In: Nicosia LM, Goldstein RA (eds) Through a distorted lens: media as curricula and pedagogy in the 21st century. SensePublishers, Rotterdam, pp 99–114

de-Marcos L, Garcia-Lopez E, Garcia-Cabot A (2016) On the effectiveness of game-like and social approaches in learning: comparing educational gaming, gamification and social networking. Comput Educ 95:99–113. https://doi.org/10.1016/j.compedu.2015.12.008

Dochy F (1994) Prior knowledge and learning. In: Husen T, Postlewaithe N (eds) International encyclopedia of education. Pergamon, London, pp 4698–4702

Dochy F, de Rijdt C, Dyck W (2002) Cognitive prerequisites and learning: how far have we progressed since Bloom? Implications for educational practice and teaching. Active Learn High Educ 3:265–284. https://doi.org/10.1177/1469787402003003006

Dörk M, Feng P, Collins C, Carpendale S (2013) Critical InfoVis: exploring the politics of visualization. In: CHI '13 Extended Abstracts on Human Factors in Computing Systems. Association for Computing Machinery, New York, pp 2189–2198

Figueira-Sampaio AdS, dos Santos EEF, Carrijo GA (2009) A constructivist computational tool to assist in learning primary school mathematical equations. Comput Educ 53:484–492. https://doi.org/10.1016/j.compedu.2009.03.012

Foster ME, Anthony JL, Clements DH, Sarama J, Williams JM (2016) Improving mathematics learning of kindergarten students through computer-assisted instruction. J Res Math Educ 47:206–232. https://doi.org/10.5951/jresematheduc.47.3.0206

Gouli E, Papanikolaou K, Grigoriadou M (2002) Personalizing assessment in adaptive educational hypermedia systems. In: De Bra P, Brusilovsky P, Conejo R (eds) Adaptive hypermedia and adaptive web-based systems. Springer, Berlin, pp 153–163

Grund CK (2015) How games and game elements facilitate learning and motivation: a literature review. In: INFORMATIK. Gesellschaft für Informatik e.V., Bonn, pp 1279–1293

Hailikari T, Nevgi A, Lindblom-Ylänne S (2007) Exploring alternative ways of assessing prior knowledge, its components and their relation to student achievement: a mathematics based case study. Stud Educ Eval 33:320–337. https://doi.org/10.1016/j.stueduc.2007.07.007

Hambleton RK, Swaminathan H (1985) Item response theory: principles and applications. Springer, Netherlands

Hanus M, Havelková L (2019) Teachers' concepts of map-skill development. J Geogr 118:101–116

Hanus M, Marada M (2016) What does a map-skills-test tell us about Czech pupils? Geografie 121:279–299. https://doi.org/10.37040/geografie2016121020279

Havelková L, Hanus M (2018) The impact of the map type on the level of student map skills. Cartographica 53:149–170

Hsiao H-S, Lin C-Y, Chen J, Peng Y-F (2018) The influence of a mathematics problem-solving training system on first-year middle school students. EurAsia J Math Sci Tech Educ 14:77–93

Hwang G-J, Sung H-Y, Hung C-M, Huang I (2013) A learning style perspective to investigate the necessity of developing adaptive learning systems. J Educ Tech Soc 16:188–197

Kimerling AJ, Buckley AR, Muehrcke PC, Muehrcke JO (2016) Map use: reading, analysis, interpretation. ESRI Press Academic, Redlands

Kitchin R, Dodge M (2007) Rethinking maps. Progr Hum Geogr 31:1–14. https://doi.org/10.1177/0309132507077082

Kosminsky D, Walny J, Vermeulen J, Knudsen S, Willett W, Carpendale S (2019) Belief at first sight: data visualization and the rationalization of seeing. Inform Des J 25:43–55. https://doi.org/10.1075/idj.25.1.04kos

Lane R, Bourke T (2019) Assessment in geography education: a systematic review. Int Res Geogr Environ Educ 28:22–36. https://doi.org/10.1080/10382046.2017.1385348

Lau WWF, Yuen AHK (2010) Promoting conceptual change of learning sorting algorithm through the diagnosis of mental models: the effects of gender and learning styles. Comput Educ 54:275–288. https://doi.org/10.1016/j.compedu.2009.08.014

Leighton JP, Gierl MJ, Hunka SM (2004) The attribute hierarchy method for cognitive assessment: a variation on Tatsuoka's rule-space approach. J Educ Meas 41:205–237

Lynch D, Howlin C (2014) Real world usage of an adaptive testing algorithm to uncover latent knowledge. In: ICERI2014 Proceedings, pp 504–511

Martín S, Rubio R (2009) Parallax cues in the design of graphics used in technical education to illustrate complex spatial problems. Comput Educ 53:493–503. https://doi.org/10.1016/j.compedu.2009.03.007
Nacheva-Skopalik L, Green S (2016) Intelligent adaptable e-assessment for inclusive e-learning. Int J Web Base Learn Teach Tech 11:21–34. https://doi.org/10.4018/IJWLTT.2016010102
Nightingale P, Te Wiata I, Toohey S, Ryan G, Hughes C, Magin D (1996) Assessing learning in universities. Professional Development Centre University of New South Wales, Sydney
Raman R, Nedungadi P (2010) Performance improvements in schools with adaptive learning and assessment. In: 2010 4th International Conference on Distance Learning and Education, pp 10–14
Schlemper MB, Stewart VC, Shetty S, Czajkowski K (2018) Including students' geographies in geography education: spatial narratives, citizen mapping, and social justice. Theor Res Soc Educ 46(4):603–641. https://doi.org/10.1080/00933104.2018.1427164
Sternberg RJ, Grigorenko (2001) All testing is dynamic testing. Issues Educ 7(2):137–170
Ting M-Y, Kuo B-C (2016) A knowledge-structure-based adaptive dynamic assessment system for calculus learning. J Comp Assist Learn 32:105–119. https://doi.org/10.1111/jcal.12119
van der Linden WJ, Glas GAW (2000) Computerized adaptive testing: theory and practice. Kluwer Academic Publishers, New York
van Dijk H, van der Schee J, Trimp H, van der Zijpp T (1994) Map skills and geographical knowledge. Int Res Geogr Environ Educ 3:68–80
Vie J-J, Popineau F, Bruillard É, Bourda Y (2017) A review of recent advances in adaptive assessment. In: Peña-Ayala A (ed) Learning analytics: fundaments, applications, and trends: a view of the current state of the art to enhance e-Learning. Springer International Publishing, Cham, pp 113–142
Wang T-H (2010) Web-based dynamic assessment: taking assessment as teaching and learning strategy for improving students' e-Learning effectiveness. Comput Educ 54:1157–1166. https://doi.org/10.1016/j.compedu.2009.11.001
Wang T-H (2014) Developing an assessment-centered e-Learning system for improving student learning effectiveness. Comput Educ 73:189–203. https://doi.org/10.1016/j.compedu.2013.12.002
Wiegand P (2006) Learning and teaching with maps. Routledge, New York
Willcockson M (1944) Some misconceptions of fifth grade children in geography. J Geogr 43(6):229–235. https://doi.org/10.1080/00221344408986988
Wongwatkit C, Srisawasdi N, Hwang G-J, Panjaburee P (2017) Influence of an integrated learning diagnosis and formative assessment-based personalized web learning approach on students learning performances and perceptions. Interact Learn Environ 25(7):889–903. https://doi.org/10.1080/10494820.2016.1224255
Wood D, Bruner JS, Ross G (1976) The role of tutoring in problem solving. J Child Psychol Psychiatr 17:89–100. https://doi.org/10.1111/j.1469-7610.1976.tb00381.x
Wu L-J, Chang K-E (2020) Effect of embedding a cognitive diagnosis into the adaptive dynamic assessment of spatial geometry learning. Interact Learn Environ. https://doi.org/10.1080/10494820.2020.1815216
Young JE (1994) Reexamining the role of maps in geographic education: images, analysis, and evaluation. Cartographic Perspect 1:10–20

Martin Hanus is an Assistant Professor at Charles University in Czechia. His research activities comprise a range of geography education issues (curriculum, teaching strategies, spatial thinking, Geocapabilites, world-mindedness). However, he dominantly focuses on the educational aspects of map use, particularly teachers' concepts of map skills development, student's misconceptions, students' strategies of map use, and maps as a tool for geographical thinking development. He teaches in the pre-service teacher-training programmes and life-long courses for in-service teachers. He is the head of the Geography Teachers' Network in Prague and a member of the Steering Committee of International Geographical Union—Commission on Geographical Education.

Lenka Havelková is an Assistant Professor at Charles University in Czechia. Her teaching and research activities are centred around educational cartography. She has developed a particular research interest in students' strategies for solving problem tasks with maps and misconceptions that influence students' understanding and use of maps. Moreover, she enjoys exploring and employing various methodological approaches from which geographical education research can benefit, such as eye-tracking experiments and conceptual tests. She translates the results of her research into practice, mainly during courses for pre-service geography teachers and when developing methodological guidelines and good-practice papers for in-service teachers.

Chapter 10
Engaging with Conceptual Acrobatics: Geography Assessment Standards and Rubrics

Deborah Heck

Abstract This chapter provides a theoretical and critical exploration of how different discourses of assessment standards relate to teacher use of criteria and standards-based rubrics in geography classrooms. Fundamentally, this work responds to concerns raised in the higher education literature that the critique of standards can be lost in the context of accountability and audit. Geography has a privileged place in the school curriculum in many countries across the world; hence, this standing in the curriculum provides an opportunity and scope for educators to critique how their understandings and philosophies impact geography assessment. A social realist lens is used to explore how three philosophies of geography standards (1) representational; (2) sociocultural; and (3) sociomaterial, relate to the use of rubrics in geography assessment. The theoretical critique provides a framework for teachers to engage with 'conceptual acrobatics' to understanding the philosophies and structures that influence the development of the reconceptualization of assessment literacy in teaching practice. Every teacher has a philosophy of assessment standard that influences the way they interpret the standards and develop rubrics. The chapter concludes with questions aimed to develop teacher assessment literacy by considering how their assessment philosophy relates to their interpretation of assessment standards and rubrics. These questions support teachers in engaging in conceptual acrobatics that uncovers how teacher interpretations of standards relate to different philosophies to develop assessment literacy. Further empirical research is required to explore how geography teachers' philosophies of assessment standards are interpreted in assessment rubrics and how assessment literacy in practice might be further developed.

Keywords Rubric · Social realist · Conceptual acrobatics · Assessment standards · Assessment literacy · Teacher philosophy

---

D. Heck (✉)
University of the Sunshine Coast, Sunshine Coast, Australia
e-mail: dheck@usc.edu.au

T. Bourke et al. (eds.), *Assessment in Geographical Education: An International Perspective*, Key Challenges in Geography,
https://doi.org/10.1007/978-3-030-95139-9_10

10.1 Introduction

National governments and international agencies are concerned with the quality of education systems with many developing systematic approaches to measure performance and outcomes. The shift towards results-based accountability requires data often on a global scale such as that provided by PISA testing, administered by the Organisation for Economic Cooperation and Development (OECD). Large-scale testing outcomes result in national comparisons that impact national education policy (Hardy 2015). In response, many nations develop standards-based reforms that directly respond to areas of concern evidenced in student performance (Bolden and Tymms 2020). Concern about the quality of teachers and teacher education has increased emphasis on the quality of teachers and teacher education programmes (Lewis et al. 2020). Education governance is increasingly concerned with data-driven approaches to policy and practice that focus on a 'what works' agenda. Hence, there is a focus on responses that can be measured (Hardy 2015) and respond to the need for education systems to be competitive and market-driven (Verger et al. 2019). Hence, in many countries, national standards have been advocated for and developed by educators and geographers. 'Geography for Life' in the USA is one example of geographers and educators advocating and positioning geography within the school curriculum (Bednarz et al. 2014). The development of these standards across the geography and geography education community significantly influenced the development of geography curriculum standards across the USA (Bednarz et al. 2014). A suite of 'Road Map' papers for geography education has been developed to maintain and develop the position and status of geography in the school curriculum (Bednarz et al. 2013; Edelson et al. 2013; Schell et al. 2013).

One way that geography standards are interpreted in teacher practice is through the use of rubrics in assessment. Lane and Bourke (2019) identify a lack of research focused on assessment in geography. Hence, this chapter's critical analysis draws on research and theory from higher education rubric use and applies this theory to teacher assessment literacy in the context of geography. Xu and Brown (2016) call for a reconceptualization of assessment literacy in teacher practice, focusing on all teachers engaging in an ongoing process to develop their assessment literacy. The Teacher Assessment Literacy in Practice (TALiP) framework developed by Xu and Brown (2016) identified the importance of teachers developing six aspects of assessment literacy. These six aspects include their assessment knowledge base, exploring conceptions of assessment, analyzing the institutional context of assessment, interpreting assessment guidelines, participating in assessment practices, and reflecting on what they have learnt to engage in the ongoing redevelopment of teacher assessor identity. This chapter also contributes to a systematic analysis of the way standards operate in specific contexts (Ajjawi and Bearman 2018). The development of this understanding contributes towards what Wheelahan (2012) suggests as engagement with theoretical knowledge, in this case about geography assessment rubrics as a representation of assessment standards. Engagement in these conversations provides opportunities to examine what might be considered not-yet-thought-about and provides a means to disrupt current practices and power structures. The dialogue aligns with the geographical inquiry process providing a means to understand fundamental knowledge and to apply that knowledge to answer questions that we have about the world around us (Favier 2011). Hence, this chapter aims to provide geography educators with access to theoretical knowledge about the connection between different views of standards and how they influence geography teachers' interpretation of standards in rubrics. It is hoped that by engaging with theoretical knowledge about standards and rubrics, how educators think about and interpret assessment practices in geography education will change and contribute toward the development of teacher professional identity (Mulcahy 2006) and assessment literacy.

The chapter begins with an overview of literature outlining types of assessment rubrics commonly used in geography. This first section also explores the impact of the 14 design decisions resulting from a range of rubrics types (Dawson 2017). The next section begins the theoretical analysis of different discourses of assessment standards in geography and their application to geography classroom rubrics. The final section explores implications for research and practice to support teachers in engaging in ongoing conceptual acrobatics to ensure that they continually (re)examine how their philosophy of assessment standards impacts the development of rubrics in the geography classroom.

10.2 Assessment Rubrics

Geography standards aim to articulate what students need to know and do. Assessment rubrics have been one mechanism adopted by educators to capture geographical performance for formative and summative purposes. Rubrics developed to provide formative feedback typically provide statements of student work quality for components of the task at standard, below standard, and above standard. Formative rubrics aim to provide information that will inform instruction and is often referred to as 'assessment for learning' (Wiliam 2011). Formative use of rubrics as feedback to revise assessment before summative assessment was identified by Lipnevich et al. (2014) to provide the largest gains in student writing performance. The study compared the growth in performance for three groups of students who were provided with formative writing support in the form of: (1) rubric only, (2) task exemplars at three standards, and (3) both the rubric and task exemplars. The study identified that the rubric-only group focused on understanding what the task required, a process they describe as 'mindfulness' that was not evidenced in the work of the other two groups.

Different approaches to grading using rubrics fall into two groups: holistic and analytic rubrics (Jonsson and Svingby 2007; Mertler 2000). Holistic rubrics are used to grade a task in accordance with the standard. Figure 10.1 represents the format for a holistic rubric. A holistic rubric is often used to assess performance or when the assessment task has no specific correct answer. In this context, an overall level

Standard A	Standard B	Standard C	Standard D	Standard E
Description of content and skills demonstrated at this standard presented as statement of understanding, proficiency, or quality.	Description of content and skills demonstrated at this standard presented as statement of understanding, proficiency, or quality.	Description of content and skills demonstrated at this standard presented as statement of understanding, proficiency, or quality.	Description of content and skills demonstrated at this standard presented as statement of understanding, proficiency, or quality.	Description of content and skills demonstrated at this standard presented as statement of understanding, proficiency, or quality.

Fig. 10.1 Holistic rubric

of accomplishment with little feedback is provided (Bennett 2016; Dawson 2017; Gezie et al. 2012).

Analytic rubrics identify components of the task that can be scored individually to generate the task's overall standard. An analytic rubric is defined by Reddy and Andrade (2010) as comprising three key components: (1) a list of criterion that will be used to evaluate the task, (2) a statement that illustrates different levels of quality for each criterion, and (3) a scoring mechanism. Analytic rubrics are represented in two typical formats—matrix or continua model (Grainger and Weir 2016). Figure 10.2 illustrates the matrix model that is typically presented in a grid format. The first column lists each criterion to be assessed, including weighting details if relevant. The remaining columns represent each performance standard, e.g., A–E (Mertler 2000). Figure 10.3 represents the continua model. The continua rubric is built with the first column representing the range of standards and the remaining columns the specified criterion. Each criterion represents the quality of the work along a continuum. The quality of student work is defined at the top and bottom of each criterion. Additional statements are only added within the continuum to illustrate how the quality of work changes for the standards identified.

These examples are just a few of the literature definitions but there remains a great deal of debate in this space (Panadero and Jonsson 2013; Reddy and Andrade 2010). In a bid to further define the complexity of rubrics, Dawson (2017) identified 14 design elements for rubrics. Dawson (2017) suggests that each element is a decision point for educators including decisions about who the rubric is shared with, the expertise of the rubric user, and the level of feedback and instruction provided with the rubric. Figure 10.4 illustrates Dawson's (2017) 14 design elements representing the critical differences between diverse approaches to rubric use in educational contexts. Xu and Brown (2016) suggest that teachers face several challenges in developing their assessment literacy. One of these challenges includes developing an understanding of how teacher conceptions of assessment and use of standards impact the development of assessment practices in schools and classrooms. The chapter will now apply this overview of rubrics to different discourses of standards and rubrics in the geography classroom.

10.3 Discourses of Assessment Standards—Rubrics in the Geography Classroom

Assessment standards in geography are part of the curriculum landscape. However, different assessment standard discourses have implications for the way teachers interpret rubrics in geography classrooms. Three discourses of assessment standards can be identified from the higher education literature: (1) representational, (2) sociocultural, and (3) sociomaterial (Ajjawi et al. 2019). This section begins the conceptual acrobatics with an overview of the three discourses of assessment standards (Ajjawi and Bearman 2018), and engages with a critical analysis of each that articulates how

Matrix Model	Standard A	Standard B	Standard C	Standard D	Standard E
Criterion 1	Quality evidenced in performance at this standard	Quality evidenced in performance at this standard	Quality evidenced in performance at this standard	Quality evidenced in performance at this standard	Quality evidenced in performance at this standard
Criterion 2	Quality evidenced in performance at this standard	Quality evidenced in performance at this standard	Quality evidenced in performance at this standard	Quality evidenced in performance at this standard	Quality evidenced in performance at this standard
Criterion 3	Quality evidenced in performance at this standard	Quality evidenced in performance at this standard	Quality evidenced in performance at this standard	Quality evidenced in performance at this standard	Quality evidenced in performance at this standard

Fig. 10.2 Matrix model format

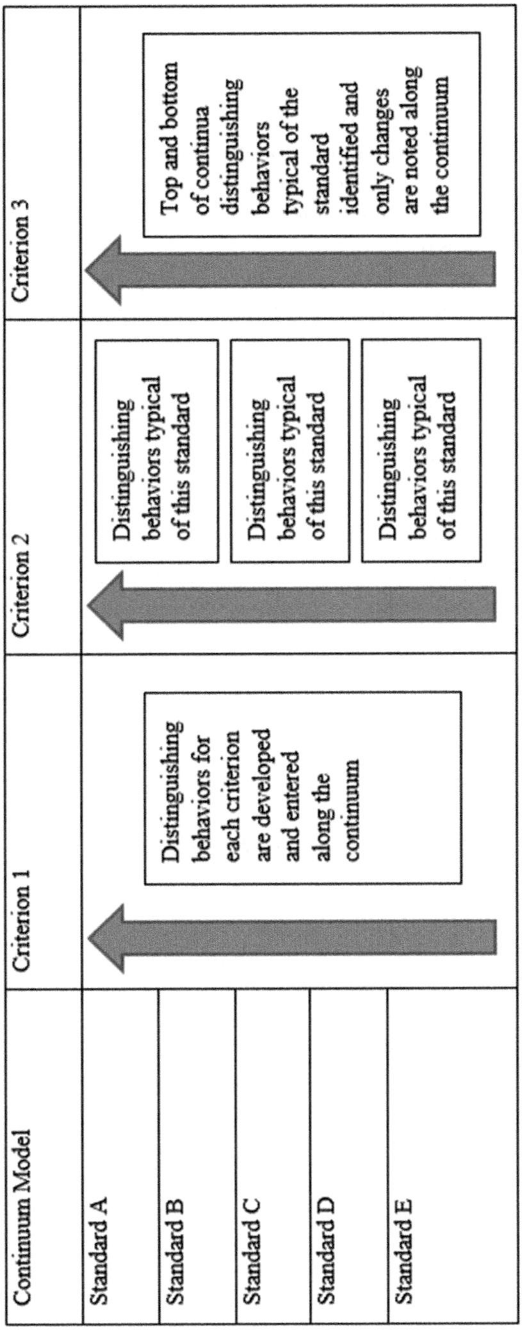

Fig. 10.3 Continuum model format

1.	Specificity: the particular object of assessment
2.	Secrecy: who the rubric is shared with, and when it is shared
3.	Exemplars: work samples provided to illustrate quality
4.	Scoring strategy: procedures used to arrive at marks and grades
5.	Evaluative criteria: overall attributes required of the student
6.	Quality levels: the number and type of levels of quality
7.	Quality definitions: explanations of attributes of different levels of quality
8.	Judgement complexity: the evaluative expertise required of users of the rubric
9.	Users and uses: who makes use of the rubric, and to what end
10.	Creators: the designers of the rubric
11.	Quality processes: approaches to ensure the reliability and validity of the rubric
12.	Accompanying feedback information: comments, annotation, or other notes on student performance
13.	Presentation: how the information in the rubric is displayed
14.	Explanation: instructions or other additional information provided to users

Fig. 10.4 Dawson's 14 rubric design elements (Dawson 2017)

teachers might interpret these discourses in geography assessment and rubrics. The analysis draws on examples from the Australian context to illustrate how different approaches to standards and rubrics in geography influence assessment practice. Exploring the arguments and justifications for rubric use and practices across the three discourses provides scope for geography educators to engage with debates and further develop their assessment literacy and identity as geography educators (Xu and Brown 2016).

10.3.1 Representational Discourse of Standards

The representational discourse identifies standards as enduring and strong—typically designed and developed by experts. The value of standards is identified through their clarity and transparency about what is being assessed (Ajjawi et al. 2019). Bloxham and Boyd (2012) identify this representational view as techno-rationalist or a positivist view of standards. They suggest that the process of assessing student work by explicit standards statements can be undertaken without the influence of power, culture, or values. The representational perspective is embedded with a view that transparency in assessment is both neutral and achievable (Bearman and Ajjawi 2018).

The dominant view of standards in geography education is the representational view. The privileged view resulted in the development of a range of explicit statements of standards for school geography. For example, in the USA, each state developed a set of standards when geography was designated a core academic subject in the early 2000s (Wertheim et al. 2013). During a similar timeframe as part of the Australian

Curriculum, geography standards of achievement were identified for each year level of compulsory schooling within the Humanities and Social Sciences learning area (Australian Curriculum, Assessment and Reporting Authority [ACARA] 2020). In this discourse, the transparency agenda contributes to the call for standards to provide explicit detail of what is expected to increase student performance (Panadero and Jonsson 2013). According to this discourse, the challenge of standards being interpreted differently by people across contexts remains (Ajjawi et al. 2019). The sociocultural discourse (see Sect. 10.3.2) seeks to address the differences in interpretation evidenced in the representational discourse's transparency claim.

10.3.1.1 Rubrics as Reality—Representation Discourse

Experts typically generate standards based on a representational discourse for use by teachers. For example, the Australian Curriculum identifies geography standards of achievement in the compulsory years of schooling from Foundation to Year 10 (aged 14–15 years) (ACARA 2019a, b). The curriculum was generated through a six-step process that began in 2009 with (1) the development of the curriculum blueprint, (2) curriculum drafts developed by expert writing teams for each learning area, (3) trial of the curriculum drafts in schools, (4) approval by the ACARA board and education ministers, (5) publication and implementations across different states and territories, and (6) annual monitoring of curriculum effectiveness (ACARA 2012). One of the curriculum development process's critical features was the identification of achievement standards for each year level and learning area. Annotated samples of student work—at and above the identified achievement standard—are presented as exemplars for teachers and students. For each year level, a range of sample assessment tasks is provided that assess different aspects of the year level achievement standard. For example, a task in Year 5 (aged 9–10 years) requires students to identify the effects of interconnection between people and the environment by analyzing three potential sites for a new hotel. Students then recommend one site based on their analysis (ACARA n.d.). In this instance, the achievement standard is identified as a holistic statement of student achievement, and the exemplars provide evidence of the achievement standard.

While a nationally developed Australian Curriculum for Geography was generated, it has been implemented differently across each state and territory where the responsibility for curriculum development is vested. Hence, the development of achievement standards and content descriptors in this context is a response by the federal government to provide organization when their direct ability to implement standards is limited (Timmermans and Epstein 2010). For example, for the state of Queensland, the responsibility for implementing curriculum has been invested in the Queensland Curriculum and Assessment Authority (QCAA). Queensland opted to implement the Australian Curriculum as published by the Australian Curriculum, Assessment and Reporting Authority (ACARA). Each state and territory in Australia makes decisions about implementation for their jurisdiction.

One of the challenges for teachers in implementing the Australian Curriculum was assessing student learning using the nationally agreed A–E levels of quality for each learning area. The C standards in the Australian Curriculum have been identified as a satisfactory achievement of the year level standard. While examples of exceeding the standard were provided, teachers sought out further details about the expected quality descriptor for the quality of the A, B, D, and E standards. In Queensland, standards elaborations have been developed to support teachers to make judgements about student work with an emphasis on consistency and comparability of judgements for work across a year (QCAA 2019a, b). The standards elaborations take the form of a rubric. The rubric begins with the inclusion of the knowledge and skills identified from the Australian Curriculum standards statement for the year level at C standard. It then uses a range of terms to elaborate on how student work quality differs across different aspects of the achievement standard. For example, the Year 7 (aged 11–12 years) standards elaboration includes a requirement to 'describe geographical processes that influence the characteristics of places' (QCAA 2019b) as evidence of achieving a C or satisfactory standard. A series of terms is used to differentiate the 'discernible differences' across achievement A–E levels. Standards above satisfactory use terms such as 'detailed' for level B and 'thorough' for level A to identify the discernible differences. In contrast, standards below satisfactory use terms such as 'partial' for level D and 'fragmented' for level E. Use of the standards elaborations provided without adaptation is an example of a representational interpretation of rubrics for assessment.

A matrix style rubric is used to present the standards elaborations. The standards elaborations contribute towards the implementation of the consistency and transparency agenda in school assessment. These publicly available standards frameworks provide a means for teachers to judge student achievement across the identified standards. It is important to note that these standards frameworks have not been developed for individual assessment tasks. Instead, they have been developed to guide judgements for assessing a portfolio of work and reporting a grade of A–E. The points of differentiation or discernible differences align with a hierarchical development in understanding. Entwistle and Smith (2002) identify five levels: (1) mentioning, (2) describing, (3) relating, (4) explaining, and (5) conceiving. The E standard represents 'mentioning' and suggests a response that lacks structure and provides fragments of information. Standard C is an example of a brief 'description' while standard D represents a mixture of 'mentioning' and 'describing'. Standards B provides an opportunity for more detailed supporting arguments to be provided demonstrating 'relating'. In comparison, standard A leaves scope for 'explaining' and 'conceiving' more independent thoughts, reflections, and arguments about the task that demonstrate higher understanding levels. The standard levels and use of discernible differences can be aligned with a range of different taxonomies (Anderson et al. 2001; Biggs and Collis 1989; Bloom et al. 1956; Marzano and Kendall 2007).

The representational discourse identified rubrics as supporting reliability in the scoring of assessment tasks (Jonsson and Svingby 2007). The discourse justifies the use of rubrics to score assessment tasks as a tool to achieve consistency of teacher judgement. The explicit identification of task requirements in rubrics based on a representational discourse focus on making the task requirements clear and transparent for students and assessors. Bearman and Ajjawi (2018) suggest that transparency allows a student to 'see through' the statements and identify the teachers' interpretation of the standard. One criticism of the use of rubrics is that students use the rubric as a checklist to generate instrumental responses that include all the aspects listed within the rubric (Sadler 2014; Torrance 2007).

10.3.2 Sociocultural Discourse of Standards

The sociocultural discourse has a less durable view of assessment standards because it rejects the notion that assessors' tacit knowledge can be made explicit (Ajjawi et al. 2019). While it is acknowledged that experts can discuss and reach a consensus about assessment standards, multiple viewpoints or perspectives about the standard are acknowledged (Ajjawi and Bearman 2018). Social, cultural, and political aspects of contexts impact assessment standards and practice, hence while standards can be identified, ongoing refinement is a requirement (Ajjawi et al. 2019).

Examples of sociocultural practices are evidenced in moderation processes where teachers engage in discussions about their interpretation of student work. There is an acknowledgement that the assessment of quality is subjective (Ajjawi et al. 2019). The sociocultural discourse advocates that assessment standards can be identified and agreed by experts. Hence, teachers and students aim to achieve the same understanding of the assessment standard (Ajjawi et al. 2019). Two challenges are associated with this discourse. First, is the lack of acknowledgement of student co-creation of rubrics in the assessment process. This results in an acceptance of the expert view identifying the second limitation that standards are not considered to be more dynamic and adaptable (Ajjawi and Bearman 2018; Ajjawi et al. 2019).

10.3.2.1 Rubrics and Context: Sociocultural Discourse

A feature of the sociocultural discourse and assessment standards is the requirement for ongoing review and refinement of standards based on assessing the impact of social, political, and cultural aspects of context. While there might be the acknowledgement of an agreed set of assessment standards, the way these are interpreted will differ based on the context, hence standards in this discourse cannot be considered neutral (Bacchus et al. 2020). One justification for using rubrics is to provide students with a context-specific opportunity to come to the same understanding as the teacher about the requirements of the task and the standard being assessed (Andrade and Du 2005). The shared understanding is achieved by using the rubric to identify the task's standards by engaging students in explaining and discussing the meaning of standards to clarify expectations and goals (Andrade 2005; Andrade et al. 2008; Brookhart and Chen 2015). Hence, in this view of standards there is a strong emphasis on teaching with rubrics.

Bijsterbosch et al. (2018) illustrate one small group example of teaching with rubrics based on pre-vocational geography teachers' work during a six-week professional development programme. The programme aimed to develop the teacher's ability to generate a school-based assessment of 'evaluating' and 'creating', including scoring rubrics based on Xu and Brown's (2016) Teacher Assessment Literacy in Practice (TALiP) framework. Teachers and researchers collaborated to explore the impact of beliefs and context on developing school-based examination questions and explored ways to generate assessment questions and rubrics to assess higher-order thinking in geography. The study identified that while teachers indicated marking using the rubric was more time consuming, they did suggest that their confidence increased when using a rubric. The confidence was based on a shared understanding of standards related to teaching and assessing student work developed through discussions. Teachers used evaluation and creation more in their everyday classroom practice, resulting in increased teacher confidence to mark student work. In this case, the assessment and marking of 'evaluating' and 'creating' in student work supported teacher assessment literacy development. The integration into everyday practice evidenced the sociocultural discourse as a means to work with students to explore interpretations of assessment standards and rubrics.

Another example of rubrics in teaching is engaging students with peer or self-assessment of their own work using rubrics. Hafner and Hafner's (2003) large scale quantitative study of college biology student peer assessment of presentations suggested that when using a rubric with a performance statement for a range of well-articulated criterion, students and academic staff provided similar assessment ratings. Sadler and Good (2006) identified that student self-assessment using rubrics led to improved performance on assessment tasks, while peer-grading did not support improvement. Use of peer or self-assessment is a formative application of rubrics in the classroom, and another example of sharing interpretations of rubrics with students as teaching. Panadero and Jonsson's (2013) review of the literature suggests that self-assessment using a rubric supports student performance in terms of: (1) understanding the task, (2) reducing assessment anxiety, (3) provision of feedback, (4) student self-efficacy, and (5) self-regulation. One of the criticisms of the use of rubrics for self-assessment relates to student use of the rubric, focusing on complying with the listed criteria (Torrance 2007). It is possible to identify rubrics focused on task-specific requirements; however, these represent a more representational discourse approach to rubrics and assessment.

A sociocultural view of standards supports teacher collaboration to develop assessment tasks that allow students to demonstrate their skills and abilities according to the achievement standards. In undertaking this work, students and teachers discuss, develop, review, and refine their understanding of the standard and how it can be assessed. Rubrics are developed for specific tasks and include descriptions of the quality of work rather than identifying specific requirements of the task (Brookhart et al. 2016). There may be multiple opportunities across the year to engage with different aspects of the required standard. Hence, the teacher makes an on-balance judgement of student achievement of the yearly standard based on the review of a portfolio of tasks for a specific learning area.

A sociocultural interpretation of the QCAA standards elaboration includes developing a suite of tasks with task-specific rubrics across the year that align with the relevant learning area curriculum requirements. Teachers then review the performances across the portfolio of tasks to assess each student's geography standard for the year. Using rubrics across a portfolio of student work is also evident in assessing geography majors at the university level (Mossa 2014). These approaches draw on a sociocultural view of the use of rubrics as context and task-specific. However, a vital criticism of the sociocultural discourse approach to interpreting rubrics is that the shared understanding or interpretation of the rubric that students need to arrive at is the teachers' interpretation of the standard. Hence, there is limited scope for co-production with students; the teacher view is privileged (Ajjawi et al. 2019).

10.3.3 Sociomaterial Discourse of Standards

The sociomaterial discourse identifies standards as coordinated in time and space while at the same time, much more dynamic and transient (Ajjawi and Bearman 2018; Ajjawi et al. 2019). In this view, standards do the work of providing order from a distance while at the same time providing scope for local decision making and adjustment during the assessment process. During assessment, it is difficult to separate the entanglement between the material, social, and individual. The assessment standards represent an artefact that does constrain or enable human action; however, each enactment is a unique response or action in its own right (Ajjawi et al. 2019).

One of the sociomaterial discourse affordances is the opportunity for 'tinkering' with the interpretation of the standards, including co-creation with students. A novice brings new ideas that provide opportunities to imagine new ways or approaches that still provide enough connection to the standard, creating a much more dynamic approach to interpreting the standards (Ajjawi et al. 2019). This discourse provides both students and assessors with an opportunity to tinker with their interpretation of the standard providing scope for generating and responding to novel and unique answers. These approaches provide scope for what Fenwick (2010) describes as a means to undo standards.

10.3.3.1 Rubrics, Tinkering, and Co-Construction: Sociomaterial Discourse

The sociomaterial discourse of standards provides a significant opportunity for teachers working in specific contexts to tinker with interpretations of the standards in collaboration with students and colleagues. It offers the opportunity to explore moments of rupture. Rupture explains the moments when the tinkering has gone too far, and the standard is no longer recognisable. It is essential for teachers and students to explore these spaces and reorient as necessary with more tinkering. Ajjawi et al. (2019) suggest that this notion of tinkering represents an activity where people participate together to generate a response that is a unique interpretation of the standard. The process represents a move towards what Ryan and Bourke (2013) identify as a 'reflexive professional' rather than a teacher who implements standards as a checklist to be completed. The sociomaterial discourse also offers students the same opportunity for reflexivity.

One example of a sociomaterial discourse of standards is students' co-construction of rubrics. The discussion and dialogue between students, teachers, and assessors about the task, rubric, and standards create an opportunity for students to develop an emotional engagement with the task and interpret the standards (Bacchus et al. 2020). The work of Fraile et al. (2017) suggests that the main benefit of engaging students in the process of developing a rubric was focused on the student's ability to internalize and develop an understanding of the assessment criteria and standards. The sociomaterial discourse recognizes the process of rubric co-construction as part of a collective social process that results in establishing a rubric as an artefact generated in a specific time and place. The rubric's joint construction provides an opportunity for participants to engage with and think about the criteria and quality through a process of social construction or cogeneration (Heck et al. 2019). The process of social construction represents tinkering, and the discussion provides opportunities for identifying ruptures that explore the boundaries. Cogeneration of rubrics in this discourse provides the prospect to illustrate to students the dynamic rather than fixed nature of knowledge (Bearman and Ajjawi 2019). It provides the freedom for students and teachers to imagine unique ways to 'see with' the criteria in open-ended ways rather than narrowing the view and 'seeing through' the standards that are evident in other discourses of standards (Bearman and Ajjawi 2018). The dynamic and iterative approach to rubric development supports teacher assessment literacy in accordance with Xu and Brown's (2016) TALiP framework.

10.4 Implications and Conclusions

The purpose of this chapter was to provide geography educators with a means to engage reflexively with the impact of assessment standards on the interpretation of rubrics in geography assessment. It offered scope for teachers to engage with the ongoing development and redevelopment of their assessment literacy in practice whether they are pre-service teachers, experienced teachers, or teacher educators, therefore contributing to the reconceptualization of teacher assessment literacy (Xu and Brown 2016). Reconceptualization occurs when teachers develop their knowledge about rubrics and different types of rubrics in geography. They can identify that different discourses of geography standards provide a means for teachers to interpret curriculum and rubrics. Teachers consider how their own conceptions of assessment influence their decision making and engagement with students. Hence, the chapter analyzed three philosophies of standards applied to events and experiences of interpreting rubrics in geography. The exploration of three discourses provided scope for teacher learning by opening up debate and uncovering the structures that influence the development of assessment practices in geography. The critical analysis offers teachers a means to reconstruct their own identity as an assessor in the geography classroom. The purpose of the final section was to call all geography educators and researchers to engage in an ongoing way with conceptual acrobatics about standards and their interpretation in rubrics to develop assessment literacy in geography. Further research is required to explore how teacher understanding of assessment discourses and interpretation of rubrics support teacher identity and assessment literacy.

In the context of geography education practice, each educator works in a specific context that provides a different set of boundaries for professional practice. Exploring the impact of different views of standards on the way rubrics are interpreted illustrates and explores possibilities for geography practice. The challenge for all geography educators is to consider how an understanding of their own view of standards and assessment can influence the way rubrics are developed and interpreted. Dawson (2017) suggests that the term rubric is challenging to define as there are so many iterations and variations that led him to develop the suite of 14 design or decision points to interpret rubrics. When interpreting rubrics, the diversity of options highlights the value and need for the ongoing development of teacher assessment literacy in practice (Xu and Brown 2016). The theoretical analysis evidenced in this chapter illustrates how educators' views of standards impact assessment rubrics in the classroom. Dawson's rubric decision points have been adapted into a suite of questions to support ongoing geography educator reflexivity when considering assessment rubrics (see Table 10.1). Teachers should begin the rubric construction process by reviewing the questions in Table 10.1 to critically reflect on the discourse of standards that they are using as they interpret rubrics for assessment (Xu and Brown 2016). If educators are to engage as enacted professionals, spaces to negotiate and tinker with assessment and rubric interpretation are required. The purpose of the list of questions is to provide a starting point for geography educators to continually engage with debate and discussion about the interpretation of rubrics and standards in geography assessment. Critical engagement with these questions aims to support teacher assessment literacy development and engage with the reconstruction of teacher assessor identity. Further research is required to explore how teacher assessment literacy develops. Bijsterbosch et al. (2019) also suggest that larger scale research is required to explore the development of geography teacher assessment literacy across multiple sites and contexts. Research should also explore whether tools such as the 14 rubric design questions (see Table 10.1) support the development of teacher assessment literacy and rubric interpretation.

While connecting the theoretical with practice provides a means for understanding how geography assessment standards are evidenced in our practice, they also support how we engage with others. As we tinker with assessment practices, we will encounter ruptures. These ruptures will provide opportunities for geography educators to engage in conversations with educators, parents, students, and administrations with various views of assessment standards. Our engagement with the theory of standards and their interpretation in the context of rubrics will provide us with insights into how our

Table 10.1 14 Rubric design questions

Rubric design feature	Questions to consider concerning each design feature
1. Specificity	Is the rubric task-specific or generic?
2. Public or private	Is the rubric public or only for the teacher/assessor?
3. Exemplars	Are examples of work at different standards available to support student understanding of different standards? Does this limit the scope for students to imagine unique ways of responding?
4. Scoring strategy	Does the rubric make clear what the scoring strategy is?
5. Evaluating criteria	Is the rubric holistic or analytic?
6. Quality levels	How many levels of quality are identified/required?
7. Quality definitions	How are the levels of quality defined?
8. Judgement complexity	Is theoretical knowledge required by users of the rubric to make judgements about standards?
9. Users and uses	Who will use the rubric and how? Are there multiple users (e.g., student, teachers, administrators, parents, and systems)?
10. Creators	Is the rubric created or co-created by the system, teachers, students, or administrators?
11. Quality processes	What quality processes impact the use and/or development of the rubric? For example, moderation and/or formal processes of approval of rubrics as representations of standards?
12. Accompanying feedback	Rubric as assessment, self or peer assessment, formative or summative?
13. Presentation	Is it a matrix model, continuum model, checklist, or a holistic statement?
14. Dialogue, explanation, or cogeneration	How important are dialogue and discussion between teachers and/or students about the rubric? Can students and teachers cogenerate possibilities?

colleagues, students, administrators, and parents discuss the challenges and opportunities they encounter with geography assessment. Understanding possible perspectives provides the means for us to challenge viewpoints, consider ideas, and discuss possibilities while also acknowledging and justifying the importance of our tinkering as geography education professionals. It is essential as engaged professionals that we continue these robust conversations even when they are difficult because it is the ongoing challenging of practice that influences change in geography education. Matshedisho (2020) reminds us of the importance of continually connecting with our students' view of rubric interpretation and identifying the implications and impact on teaching and assessment.

While this chapter has focused on assessment and rubrics as a space for tinkering, it also provides a call for geography educators to consider being reflexive about how different views of standards impact our work and identity as geography educators. Hence, for all geography educators, the challenge is to recognize the possible spaces to tinker in the geography classroom; to acknowledge that there will be moments of rupture where things might be stopped in their tracks by systems, processes, and people. Do not be discouraged; it is crucial to identify the boundaries of our tinkering in geography education and recognize that these can change over time. Refocus, engage in more conceptual acrobatics, discuss with others, and return to tinkering to reconceptualize assessment literacy and identity as assessors in geography.

References

Ajjawi R, Bearman M (2018) Problematising standards: representation or performance? In: Boud D, Ajjawi R, Dawson P, Tai J (eds) Developing evaluative judgement in higher education: assessment for knowing and producing quality work. Routledge, London, pp 41–50

Ajjawi R, Bearman M, Boud D (2019) Performing standards: a critical perspective on the contemporary use of standards in assessment. Teach High Educ. https://doi.org/10.1080/13562517.2019.1678579

Anderson LW, Krathwohl DR, Airasian PW, Bloom BS (2001) A taxonomy for learning, teaching, and assessing: a revision of Bloom's taxonomy of educational objectives. Longman, New York

Andrade HG (2005) Teaching with rubrics: the good, the bad, and the ugly. Coll Teach 53(1):27–30. https://doi.org/10.3200/CTCH.53.1.27-31

Andrade H, Du Y (2005) Student perspectives on rubric-referenced assessment. Practical Assess Res Eval 10(1). https://scholarworks.umass.edu/pare/vol10/iss1/3

Andrade HL, Du Y, Wang X (2008) Putting rubrics to the test: the effect of a model, criteria generation, and rubric-referenced self-assessment on elementary school students' writing. Educ Meas 27(2):3–13. https://doi.org/10.1111/j.1745-3992.2008.00118.x

Australian Curriculum, Assessment and Reporting Authority [ACARA] (n.d.) Humanities and social science: work sample year 5—hotel proposal. https://www.australiancurriculum.edu.au/resources/work-samples/samples/hotel-proposal-at/. Accessed 11 Nov 2020

Australian Curriculum, Assessment and Reporting Authority [ACARA] (2012) Curriculum development process, Version 6. https://docs.acara.edu.au/resources/ACARA_Curriculum_Development_Process_Version_6.0_-_04_April_2012_-_FINAL_COPY.pdf

Australian Curriculum, Assessment and Reporting Authority [ACARA] (2019a) 7–10 geography. http://www.australiancurriculum.edu.au/humanities-and-social-sciences/geography/curriculum/f-10?layout=1. Accessed 13 Jan 2021

Australian Curriculum, Assessment and Reporting Authority [ACARA] (2019b) Australian Curriculum: humanities and social sciences F–6. https://www.australiancurriculum.edu.au/download?view=f10. Accessed 6 Feb 2021

Australian Curriculum, Assessment and Reporting Authority [ACARA] (2020) Australian curriculum: humanities and social sciences. https://www.australiancurriculum.edu.au/f-10-curriculum/humanities-and-social-sciences/

Bacchus R, Colvin E, Knight EB, Ritter L (2020) When rubrics aren't enough: exploring exemplars and student rubric co-construction. J Curric Pedagog 17(1):48–61. https://doi.org/10.1080/15505170.2019.1627617

Bearman M, Ajjawi R (2018) From "seeing through" to "seeing with": assessment criteria and the myths of transparency. Front Educ. https://doi.org/10.3389/feduc.2018.00096

Bearman M, Ajjawi R (2019) Can a rubric do more than be transparent? Invitation as a new metaphor for assessment criteria. Stud High Educ. https://doi.org/10.1080/03075079.2019.1637842

Bednarz SW, Heffron S, Huynh NT (2013) A road map for 21st century geography education: geography education research. Association of American Geographers, Washington. http://media.nationalgeographic.org/assets/file/NGS_RoadMap_GERC_6-21.pdf

Bednarz SW, Heffron SG, Solem M (2014) Geography standards in the United States: past influences and future prospects. Int Res Geogr Environ Educ 23(1):79–89. https://doi.org/10.1080/10382046.2013.858455

Bennett C (2016) Assessment rubrics: thinking inside the boxes. Learn Teach 9(1):50–72. https://doi.org/10.3167/latiss.2016.090104

Biggs J, Collis K (1989) Towards a model of school-based curriculum development and assessment using the SOLO taxonomy. Aust J Educ 33(2):151–163. https://doi.org/10.1177/168781408903300205

Bijsterbosch E, Béneker T, Kuiper W, van der Schee J (2018) Characteristics of test items focusing on meaningful learning: a case study in pre-vocational geography education in the Netherlands. Eur J Geogr 9(1):62–79

Bijsterbosch E, Béneker T, Kuiper W, van der Schee J (2019) Teacher professional growth on assessment literacy: a case study of prevocational geography education in the Netherlands. Teach Educat 54(4):420–445

Bloom BS, Engelhart MD, Furst EJ, Hill WH, Krathwohl DR (1956) Taxonomy of educational objectives: the classification of educational goals. Handbook I: cognitive domain. David McKay Company, Ann Arbor

Bloxham S, Boyd P (2012) Accountability in grading student work: securing academic standards in a twenty-first century quality assurance context. Br Educ Res J 38(4):615–634. https://doi.org/10.1080/01411926.2011.569007

Bolden D, Tymms P (2020) Standards in education: reforms, stagnation and the need to rethink. Oxf Rev Educ. https://doi.org/10.1080/03054985.2020.1781608

Brookhart SM, Chen F (2015) The quality and effectiveness of descriptive rubrics. Educ Rev 67(3):343–368. https://doi.org/10.1080/00131911.2014.929565

Brookhart SM, Guskey TR, Bowers AJ, McMillan JH, Smith JK, Smith LF, Stevens MT, Welsh ME (2016) A century of grading research: meaning and value in the most common educational measure. Rev Educ Res 86(4):803–848. https://doi.org/10.3102/0034654316672069

Dawson P (2017) Assessment rubrics: towards clearer and more replicable design, research and practice. Assess Eval High Educ 42(3):347–360. https://doi.org/10.1080/02602938.2015.1111294

Edelson DC, Shavelson RJ, Wertheim JA (2013) A road map for 21st century geography education: assessment. National Geographic Society, Washington. http://media.nationalgeographic.org/assets/file/NGS_RoadMapConcept_AssessmentReport_06_1.pdf

Entwistle N, Smith C (2002) Personal understanding and target understanding: mapping influences on the outcomes of learning. Br J Educ Psychol 72(3):321–342. https://doi.org/10.1348/000709902320634528

Favier T (2011) Geographic Information Systems in inquiry-based secondary geography education: theory and practice. Dissertation, VU University, Amsterdam

Fenwick TJ (2010) (un)Doing standards in education with actor-network theory. J Educ Pol 25(2):117–133. https://doi.org/10.1080/02680930903314277

Fraile J, Panadero E, Pardo R (2017) Co-creating rubrics: the effects on self-regulated learning, self-efficacy and performance of establishing assessment criteria with students. Stud Educ Eval 53:69–76. https://doi.org/10.1016/j.stueduc.2017.03.003

Gezie A, Khaja K, Chang VN, Adamek ME, Johnsen MB (2012) Rubrics as a tool for learning and assessment: what do baccalaureate students think? J Teach Soc Work 32(4):421–437. https://doi.org/10.1080/08841233.2012.705240

Grainger P, Weir K (2016) An alternative grading tool for enhancing assessment practice and quality assurance in higher education. Innovat Educ Teach Int 53(1):73–83. https://doi.org/10.1080/14703297.2015.1022200
Hafner JC, Hafner PM (2003) Quantitative analysis of the rubic as an assessment tool: an empirical study of student peer-group rating. Int J Sci Educ 25(12):1509–1528
Hardy I (2015) Data, numbers and accountability: the complexity, nature and effects of data use in schools. Br J Educ Stud 63(4):467–486. https://doi.org/10.1080/00071005.2015.1066489
Heck D, Grimmett H, Willis L-D (2019) Teacher educators using cogenerative dialogue to reclaim professionalism. In: Gutierrez A, Fox J, Alexander C (eds) Professionalism and teacher education: voices from policy and practice. Springer, Singapore, pp 137–156. https://doi.org/10.1007/978-981-13-7002-1_7
Jonsson A, Svingby G (2007) The use of scoring rubrics: reliability, validity and educational consequences. Educ Res Rev 2(2):130–144. https://doi.org/10.1016/j.edurev.2007.05.002
Lane R, Bourke T (2019) Assessment in geography education: a systematic review. Int Res Geogr Environ Educ 28(1):22–36. https://doi.org/10.1080/10382046.2017.1385348
Lewis S, Savage GC, Holloway J (2020) Standards without standardisation? Assembling standards-based reforms in Australian and US schooling. J Educ Pol 35(6):737–764. https://doi.org/10.1080/02680939.2019.1636140
Lipnevich AA, McCallen LN, Miles KP, Smith JK (2014) Mind the gap! Students' use of exemplars and detailed rubrics as formative assessment. Instr Sci 42(4):539–559. https://doi.org/10.007/s1125-013-9299-9
Marzano RJ, Kendall JS (2007) The new taxonomy of educational objectives, 2nd edn. Corwin Press, Thousand Oaks
Matshedisho KR (2020) Straddling rows and columns: students' (mis)conceptions of an assessment rubric. Assess Eval High Educ 45(2):169–179. https://doi.org/10.1080/02602938.2019.1616671
Mertler CA (2000) Designing scoring rubrics for your classroom. Practical Assess Res Eval 7(1). https://doi.org/10.7275/gcy8-0w24
Mossa J (2014) Capstone portfolios and geography student learning outcomes. J Geogr High Educ 38(4):571–581. https://doi.org/10.1080/03098265.2014.958659
Mulcahy D (2006) The salience of space for pedagogy and identity in teacher education: problem-based learning as a case in point. Pedagog Cult Soc 14(1):55–69. https://doi.org/10.1080/14681360500487827
Panadero E, Jonsson A (2013) The use of scoring rubrics for formative assessment purposes revisited: a review. Educ Res Rev 9:129–144. https://doi.org/10.1016/j.edurev.2013.01.002
Queensland Curriculum & Assessment Authority [QCAA] (2019a) 7–10 geography standards elaborations. Queensland Government. https://www.qcaa.qld.edu.au/p-10/aciq/standards-elaborations/p-10-hass/p-10-geography. Accessed 11 Nov 2020
Queensland Curriculum & Assessment Authority [QCAA] (2019b) P–7 humanities and social sciences (HASS) standards elaborations. Queensland Government. https://www.qcaa.qld.edu.au/p-10/aciq/standards-elaborations/p-10-hass/p-7-hass. Accessed 11 Nov 2020
Reddy YM, Andrade H (2010) A review of rubric use in higher education. Assess Eval High Educ 35(4):435–448. https://doi.org/10.1080/02602930902862859
Ryan M, Bourke T (2013) The teacher as reflexive professional: making visible the excluded discourse in teacher standards. Discourse 34(3):411–423. https://doi.org/10.1080/01596306.2012.717193
Sadler DR (2014) The futility of attempting to codify academic achievement standards. High Educ 67(3):273–288. https://doi.org/10.1007/s10734-013-9649-1
Sadler P, Good E (2006) The impact of self- and peer-grading on student learning. Educ Assess 11(1):1–31. https://doi.org/10.1207/s15326977ea1101_1
Schell EM, Roth KJ, Mohan A (2013) A road map for 21st century geography education: Instructional materials and professional development. National Council for Geographic Education, Washington. http://media.nationalgeographic.org/assets/file/NGS_RoadMapIMPDReport_6-21.pdf

Timmermans S, Epstein S (2010) A world of standards but not a standard world: toward a sociology of standards and standardization. Annu Rev Sociol 36(1):69–89. https://doi.org/10.1146/annurev.soc.012809.102629

Torrance H (2007) Assessment as learning? How the use of explicit learning objectives, assessment criteria and feedback in post-secondary education and training can come to dominate learning. Assess Educ Princ Pol Pract 14(3):281–294. https://doi.org/10.1080/09695940701591867

Verger A, Fontdevila C, Parcerisa L (2019) Reforming governance through policy instruments: how and to what extent standards, tests and accountability in education spread worldwide. Discourse 40(2):248–270. https://doi.org/10.1080/01596306.2019.1569882

Wertheim JA, Edelson DC, The Road Map Project Assessment Committee (2013) A road map for improving geography assessment. Geogr Teach 10(1):15–21. https://doi.org/10.1080/19338341.2012.758044

Wheelahan L (2012) Why knowledge matters in curriculum: a social realist argument. Routledge, London. https://doi.org/10.4324/9780203860236

Wiliam D (2011) What is assessment for learning? Stud Educ Eval 37(1):3–14. https://doi.org/10.1016/j.stueduc.2011.03.001

Xu Y, Brown GTL (2016) Teacher assessment literacy in practice: a reconceptualization. Teach Teach Educ 58:149–162. https://doi.org/10.1016/j.tate.2016.05.010

Deborah Heck is an Associate Professor in the School of Education and Tertiary Access at the University of the Sunshine Coast, Australia. Her research explores education theory, policy, curriculum, assessment and practice through dialogue and contributes to understanding teaching, teachers, and teacher education. In addition, Deborah teaches secondary geography curriculum and research design coordinating the Master of Education and Research Higher Degree programs. She has received seven fellowships, teaching and research awards focussed on e-learning, team teaching, equity and diversity and engagement excellence.

Chapter 11
Reflecting on the Assessment Landscape of Primary Geography

Simon Catling

Abstract Assessment is a significant element in the planning, teaching, assessing, and reviewing cycle of classroom practice in all subjects, including geography, in primary schools. Geography is one of the many subjects taught by primary teachers, a circumstance which may well impact on their assessment of their class's geographical learning. However, while there is some expert guidance on why and how to undertake assessment in geography, almost nothing is known about what primary teachers in fact do in their geography assessment practices. Their experiences are not shared in teaching and academic geography education journals which, furthermore, report no research into what primary teachers do. Drawing on a number of guidance texts about primary geography, the humanities, and social studies, several dimensions of primary geography assessment are identified, though not to promote particular approaches and benefits or limitations. The dimensions discussed are: the purpose of geography assessment; understanding what children know about at the start and during a geography topic; planning and assessment; progression and assessment in geography; approaches and activities which teachers might use to undertake assessment; recording children's learning and assessment in curriculum; and teaching accountability. Given the lack of research, what emerges are suggested research questions to gain insight into the practices of teachers and schools, so as to effect improvements. These questions form a significant agenda for research into primary geography assessment. Four priorities are highlighted: primary teachers' intentions in their geography assessment; how their assessment practices aide children's learning in geography topics; how their assessment intentions and practices relate effectively to the geography curriculum topics they teach; and how what teachers do and learn about their children's geographical understanding enables them to improve their teaching and children's learning of geography.

Keywords Primary geography · Geography assessment · Assessment dimensions · Research agenda

---

S. Catling (✉)
School of Education, Harcourt Hill, Oxford Brookes University, Oxford, UK
e-mail: sjcatling@brookes.ac.uk

T. Bourke et al. (eds.), *Assessment in Geographical Education: An International Perspective*, Key Challenges in Geography,
https://doi.org/10.1007/978-3-030-95139-9_11

11.1 Introduction: Setting the Scene

There is much common ground across the educational spectrum about the value, roles, and practices of assessment. Shared interests in the purposes of assessment in teaching children, the relationship between curriculum intentions, expectations and assessment, the variety of assessment activities that can be used, matters of informal and formal assessment, and focuses such as assessment for learning and assessment of learning, have produced a significant literature (Gardner 2012). These elements emphasize the importance of children's learning, gains, needs, and achievements (Clarke 2014; Spendlove 2009; Christodoulou 2016). Considered alongside them are the interpretation and use of assessment data by a school's leadership team and by local and state authorities and other providers of schooling, government policy makers, and educational and other consultants and advisers (Assessment Reform Group [ARG] 2005; Alexander 2010; Gardner 2012; Dann 2014; Faragher 2014; Hayward and Hayward 2016; Peacock 2016; Briggs 2019; Pollard 2019; Chiles 2020; Donarski 2020). Assessment serves a wide variety of needs and interests from supporting children's learning to accountability.

As with teachers across the educational spectrum, primary[1] teachers use a wide variety of approaches in assessment in their classrooms. These include observation, listening, discussion, and play-based activities, extended by written, mathematical, graphic, physical, artistic, and enactive means of assessing and recording what children do. The aim is to help teachers understand a whole child's engagement, learning, achievements, behavior, needs, progress, and much more. Primary teachers focus formatively on children using assessment continuously, interactively, and interpersonally to enable them to support their children's ongoing development (Wiliam 2011) rather than focusing on summative assessments, such as tests, at given points (ARG 2005) though these are certainly used when timely and informative. Indeed, primary teachers may apply a range of approaches across subjects because they teach the full curriculum, quite probably not specialists in any of them. Their assessment practices may be influenced by required emphases on the national language and mathematics dominating their curriculum, but approaches encountered through other subjects, such as geography, social sciences, and physical education, may extend their repertoire.

School decisions may affect what primary teachers assess and how they do this. For instance, a subject-based primary curriculum may well involve them in different emphases and ways to gather information about children's learning than an interdisciplinary curriculum. The breadth of the curriculum may also be an important factor, since primary teachers may feel more knowledgeable and confident about some subjects compared to others and thus be better equipped in some subjects to judge when a child understands particular content well or is having difficulties in

[1] The word 'primary' in this chapter refers to children in school between the ages of 3/4 to 11/12. Geography, or learning about the world, may or may not be included in their curriculum across this age range. Some 4-year-olds begin to learn about the world, while other children do not formally do so until 7 or 8 years old.

appreciating key patterns or ideas. Similarly, primary teachers may find guidance about when to use and how to apply various assessments either straightforward or challenging because of their varied subject knowledge strengths and limitations. Given the range of their teaching and the expectations on them to know in detail how well the children they teach are achieving in their learning and their need to provide assessment information for the senior leadership team of their school, it may not be surprising that primary teachers undertake assessment more effectively in some aspects of their curriculum than others. These are some of the contexts which impact on geography assessment.

This chapter raises and reflects on a variety of matters and questions concerning assessment in primary geography teaching, bearing in mind that there is much to be researched and discovered about the value of assessing primary children's geographical learning. It is important to state at the outset that this chapter does not set out to suggest the most appropriate contexts in which to undertake assessment in primary geography, nor does it identify what might be the most helpful activities to use, how, when, and for what purpose. There is much advice about this but negligible evidence about what really goes on in classrooms. The focus here is to consider a range of matters concerning assessment in primary geography teaching and suggest an agenda for future research for this much under-investigated aspect of primary geography. The concern is that what is considered to be appropriate direction in the nature, value, effectiveness, and practices in primary geography assessment is not grounded in detailed investigations of what primary teachers do in their classrooms.

11.2 (Mis)Placed Assessment Research in Primary Geography

Assessment is—or should be—a key component of geographical education in primary schools, as essential as considerations about what subject matter might be included in a school's geography programme, the nature and organization of its curriculum as children encounter it through their primary schooling, the quality of its teaching, the nature of the learning environment socially and physically, and access to good and appropriate resources and external places and sites. Assessment is important because it is through the variety of ways teachers assess their children, formally and informally, that they understand, appreciate, and can build on what the children know, can do, and how they respond to and feel about their geography learning and teaching. It is important that primary teachers assess children on topics and content in ways appropriate for the subject.

Yet, primary geography assessment is a problematic arena. An appropriate description of it might be as a landscape of least interest and greatest ignorance when it comes to knowing what primary teachers do and with what impacts in geography, since it is largely unexplored. Potentially, this lets down children in primary schools, who benefit from knowing what they are achieving, what to do about what challenges them, and where they are going in their geographical learning. This is not to say that primary teachers do not assess children's geographical learning, achievements, and needs, nor use what they learn in their planning. But the situation is that outside any specific primary school or classroom, no one really knows what children are gaining or losing, and what needs to be undertaken to effect improvements in this situation. Little evidence exists about what happens in classes, even where there are efforts made to gain a sense of the nature and quality of what is happening in geography in primary schools, as in England (Ofsted 2011).

The Geographical Association's *Primary Geography* is the only journal available internationally which focuses solely on geography in primary teaching. It provides practical guidance and case studies at classroom and school levels for primary teachers to utilize. A review of articles in *Primary Geography* between 2010 and 2020 provides no reference specifically and explicitly to primary geography assessment. The range of fascinating topics and classroom approaches and practices presented in the journal might provide intentions, expectations, activities and tasks, and promote further ideas that might be used in informal and formal assessment activities, but their value for and in assessment is almost never referred to. Is it surprising, therefore, that there is negligible evidence of assessment in primary geography and that this is an area of weakness among primary teachers, since it receives so little debate, examination, and encouragement?

An examination of geography education research journals, for example, the *Journal of Geography*, *Research in Geographic Education*, and *International Research in Geographical and Environmental Education*, over the same period leads to a similar conclusion. Not only is research and discussion about assessment in geography in secondary schools limited, it is virtually non-existent about geography in primary education and schooling, as it is in much else (Catling 2013). This point is well illustrated in Lane and Bourke's (2019) wide-ranging review of assessment across geography schooling since 2000. They demonstrate that the paucity of research and debate about assessment in primary geography is a global concern! This is reinforced by the report on the state of assessment in geography provided by the *Road Map for the 21st Century Geography Education Project* in the USA (Edelson et al. 2013). Though not distinguishing elementary geography (taught within social studies curricula) assessment from geography assessment with junior high children, this report provided no insight into assessment priorities, foci, and practices with and for younger children in their classrooms. Grant's (2017) social studies review was similarly uninformative, with the research evidence very limited. Enlightenment on this matter for primary geography is equally unavailable in Butt's (2020) review of geography education research, where discussion about research on assessment in confined essentially to national assessment, be these national examinations or standardized testing (Marcinkowski et al. 2013; Grant and Salinas 2008; Grant 2017). This might indicate an end of the road for primary geography assessment; then again, it might just be the starting-point.

11.3 Assessment in Primary Geography: The Lie of the Land

The lack of a research base about assessment in primary geography has not constrained proposing thoughtful and useful guidance about this key aspect of geography's teaching and learning. Indeed, there is a variety of advice available through publications and websites about primary geography (Geographical Association [GA] 2015; Catling and Willy 2018; Barlow and Whitehouse 2019; Owens 2019), as well as on geography in the context of humanities and social sciences (Buchanan 2013; Pickford et al. 2013; Reynolds 2019; Grigg and Hughes 2019). These sources are drawn on in the following sections, but reference to them is not provided unless essential, to avoid constant repetition. The guidance provided by primary geography and other experts is concerned with a wide range of elements in geography assessment, from its purpose and focus to children's engagement and achievements and to ways to assess and record learning. This section draws on this advice and notes the variety of approaches advocated, but does not attempt to cover all the elements in geography assessment that are referred to. The assessment guidance put forward by primary geography experts is drawn from experience, not from a research base. This is not to play down their value, but it raises questions which need to be pursued to develop and improve primary geography assessment.

11.3.1 Purpose for Primary Geography Assessment

Assessment in primary geography, including within its humanities and social studies contexts, is justified for a variety of reasons. The core justification is that assessment is essential within the geography planning, teaching, and evaluation cycle because teachers need to know whether or not what they expect the children to learn has been achieved. Further reasons given are that assessment must be used to inform teachers about what relevant geography the children already know, the effect of their teaching on the children's geographical learning, how teachers understand children's progress and needs in their learning, and what they require or encourage children to do next within their geography topic or in their next topic. Furthermore, assessment is argued to be of value for the children, enabling them to have a sense of their own achievements, needs, and next steps in their geographical journey. It should help them see where they may need to revisit or follow-up aspects of geography they have found challenging as well as build on what they feel confident about. Geography assessment should support their personal target setting and what they might focus their learning on within this. In these ways, assessment can fulfil what ideally is its prime function, which is to motivate children's learning in geography. Encouraging children with positive remarks about their involvement in their activities and their achievements is noted in some guidance to be a significant stimulus to deeper and longer-term engagement with the subject, as is helpful advice and information which enables children to overcome difficulties they have and understand what they are working on and towards. Assessment in this way is an ongoing and almost informal but captivating means to maintain children on task and provide direction and value in their geographical studies.

Assessment in primary geography helps the teacher who is the lead for each subject in their school, such as the geography subject leader, to establish and monitor how classes of children are doing in relation to the topics and expectations set for that year. Indeed, it can help in determining whether these need to be adjusted to be more or less demanding in future teaching. Assessment, then, acts as a check against internal or external expectations and standards. It might be used with or across year groups to identify overall performance and changes to a school's curriculum. Thus, geography assessment is able to support curriculum monitoring and evaluation, teachers' performances in providing children's geographical learning, change and development such as of resources and the geography content and its sequence, and the accountability of the subject leader in ensuring the quality, achievements, and developments for the school's children in geography.

This is a significant range of reasons for assessing geography teaching and learning in and across a primary school, to enable within-lesson and lesson-to-lesson learning and to know periodically what children have gained, or not, from their geographical studies. From perusing primary school's geography policies and internal guidance for teachers, it seems that while assessment is often mentioned, what is stated about it is varied and at times very brief. What is particularly unclear are the reasons teachers and schools have for assessing children's geographical learning beyond the obvious brief statements outlined above. In many geography policy statements, primary schools indicate what they assess in very broad subject content terms. They may list some forms or methods they use. What is usually lacking is information and explanations about when or how and to what effect they use assessment in geography teaching. The implication might be that what they state will be readily understood among the staff, who may have access to non-public documents. Of course, brevity gives little away in place of detailed information about practices, and generalization retains internal control. Where schools provide reasons, they refer most often to checking children's geographical knowledge and skills in relation to national or local requirements and to recording the children's geographical learning; rarely do they say more than this. As such, this is an area ripe for and in need of detailed and thorough classroom and school-based investigations to know and appreciate why, what, and how teachers and schools use assessment in primary geography for the benefit of their children. The need is to investigate rigorously whether there is depth and breadth in thinking about assessing primary geography and whether and how this affects the quality of practice in teaching primary geography and children's geographical learning.

Suggested research questions to pursue
Purposes of primary geography assessment

Do primary school policy documents state that they assess children's geographical learning, achievements, and needs? Is this stated specifically for geography or in general terms about its value and their practices? Why is this, how is it expected to be used, and what is the impact?

What are primary teachers' reasons for assessing their children's geography? Why do they give these reasons? Are these reasons implemented through what teachers do in their practice?

How do primary teachers interpret the school's reasons for the assessment of geography, and how does this affect their approach to assessment? Does this benefit the children?

How do children understand the purposes of assessment in learning geography? Are they aware of being assessed, and if so how and about what? Do they perceive assessment in geography as motivational, as information giving, or in some other ways? What would they like assessment to provide for them?

11.3.2 Positioning Assessment in the Scheme of Things

The point has been made that assessment is a key aspect of the planning and teaching cycle. This needs to be considered further, not least because children do not come to any geography topic as 'blank slates', and because what they study in the short-term sits within the context of their geographical learning as they develop through their primary school, supporting progress in their understanding. Three interlinking matters, therefore, need to be considered: understanding the geographical backgrounds which children bring to their primary classes, connecting assessment with planning the geography curriculum and teaching, and progression and assessment in primary geography.

Children coming into primary settings and classes do not come without geographical awareness, experience, and knowledge whether or not they are aware of it. What is known about younger children's geographical backgrounds has been extensively documented through research reports and reviews about their geographical and environmental experiences (Matthews 1992; Chawla 2002; Freeman and Tranter 2011; Holt 2011; Griffiths 2013; Blundell 2016; Nairn and Kraftl 2016; Freeman and van Heezik 2018; Malone 2018; Scoffham 2019). While such research has been undertaken both in and outside school time, it has been external to the teaching of geography, focused on understanding children's development and their geographies. This has involved studies principally of children in and their senses of their localities and the wider world, as well as of their play and school places and spaces. On occasion the word 'assessment' has been used in an article title on an aspect of their geographies (e.g., Harwood and Rawlings 2001), though the word is a misnomer since such articles focus on research to gain insight into the range and development across an age group of children's understanding of an aspect of their geography; such studies are not an assessment of children as individuals to enable their classroom learning though they might be informative for teachers in general terms.

It is advocated that primary teachers should know about their children's geographies (Catling and Willy 2018; Barlow and Whitehouse 2019), and can do so by finding out about what children know at the start of geography topics. This is an approach using diagnostic assessment. At the start of a local place study children may be asked to draw a 'mental map' of their local area, or for a topic about a distant place they may undertake a word association task. Such activities can provide information about individual children's knowledge, perceptions, and attitudes to inform a teacher about their preconceptions pertinent to a geographical study, but it is not known how useful these tasks are in informing teachers about their children nor whether what teachers learn informs adjustments to their planning. Such diagnostic tasks can be used at points during topics to gauge children's awareness relevant to a new aspect within a topic or place study, but there is no research about this. There is a need to investigate both these areas since they can enable primary teachers to understand their children's starting points and what may affect their ongoing geographical learning.

To initiate a geography topic by assessing what the children might know that is relevant to it may be an initiating stimulus for the study or come at an early point. Such an approach recognizes that a diagnostic assessment is a valuable element in planning, also able to be used at later points to build on knowledge or to address misunderstandings. It reinforces the point that assessment needs to be considered and planned for in the teaching of a geography study. In setting expectations for the outcomes of a topic, teachers will be clear about where they intend their studies to take the children's learning and achievements. This is indicated to be a common practice, though what this practice is and how it is effective remains unknown. It may be linked with formative assessment or result in summative assessment based on the cumulative learning across the topic about specific aspects which are deemed to be the key achievements expected. Such assessment may enable teachers to record what their children have gained from their studies, not simply be a record of what they have studied. However, what primary teachers do in this context requires investigation.

Suggested research questions to pursue
Understanding children's geographies

What geographical experience, awareness, knowledge, and skills do primary children bring to topics they study in their class? How is this information sought and used by their teacher—or is it not—and how does it impact on their planning and adjustments to it?

In what ways do primary teachers, if at all, investigate their children's knowledge—its strengths and limitations—of new aspects within geographical topics being studied? How might such diagnostic practice assist primary teachers?

How do diagnostic assessments of children's geographies aid their learning about the topics they study? Which types of activities do children consider help them think diagnostically about their current knowledge?

These intentions for a topic should be held in mind by the teacher throughout their teaching and the children's learning. Indeed, a teacher may have a sense of the steps or achievements that might be taken along the children's geographical journey. Planning with a sense of helping children to progress in their learning is the basis of assessment for learning. This requires that a primary teacher be clear about the journey the geography topic may take, has this constantly in mind in order to be able to respond to children's questions, provide guidance on what to do and where to head next, and is open-minded to adjusting the topic as particular needs and misunderstandings are identified, whether by changing the pace of teaching or introducing new inputs, support, or strategies. Such monitoring of children's progress in their geographical learning is argued to be central to effective planning and teaching.

Yet, while these approaches are strongly advocated in geography guidance, how they are enacted remains very unclear. The lack of research into the role of assessment in planning and teaching primary geography means that while general principles and practices may be drawn upon, the guidance provided is based in the practice expertise of advocates, not in well-grounded evidence about what primary teachers in fact do during their teaching of geography. It remains unclear in what ways and to what extent formative and summative assessment in geography teaching benefits children and in what ways this may need to be rethought. This is not to decry either the expertise or guidance proffered, which teachers may well find helpful, but to note that the lack of investigation across teachers, classes, and schools means practices remain untested and unevaluated. This aspect of the approach to assessment is in real need of research.

Suggested research questions to pursue
Planning and assessment

How and why does considering assessment opportunities during their planning of a geography topic enhance or constrain their planning? What approaches to assessment do they plan in? What do they consider would be helpful in improving their planning with assessment in mind?

To what extent do primary teachers use formative and summative assessment in their geography teaching? What are their reasons for doing or not doing so? What impact on their approach to teaching and on children's geographical learning do they consider their decisions have?

> What assessment approaches to their geography teaching do primary teachers use, when are these applied, and how are they useful to them?

Planning for geography teaching in a geography topic, across the topics for a class and across the school age groups as children move through the school, requires the subject leader for geography and teachers of classes to have a clear sense of progression in the geography curriculum and for children's geographical learning. Clarity in topic planning should enable planning for progression within a topic. Indeed, planning and sequencing the geography topics across the year surely must be based in enabling children to make progress in their geographical understanding, not simply to cover listed content. It might be that their final geography topic for a year, while being in a new aspect of geography, draws on key concepts, skills, and values so that this topic's assessment may provide a sense of the children's cumulative learning about geography to pass on to their next teacher. However, this is only part of the picture. For this to be effective a school needs to provide an explicit sense of progression in the geographical content, topics, thinking, and learning through children's primary schooling. However, this is problematic.

In their geography policies, primary schools may or may not provide information about the sequence of what is taught in the subject. This may, misleadingly, indicate that there is progression in learning geography. Where a content sequence is provided it seems usual to set this out by age phase, rarely by year group. While this may show the range of aspects of geography studied, it provides no indication of ways in which geographical knowledge is taken forward to enhance and deepen children's understanding of the subject. Children may simply accumulate more information. Some primary schools provide broad attainment outcomes for the end of primary schooling but rarely any sense of how these are to be achieved. A few schools may illustrate ways in which they provide progression in an aspect of geography, such as about mapwork skills, though again this may be little more than the sequence in their year planning. This is not surprising, since guidance about planning for progression in primary geography tends to be broad-brush rather than to give detail about how a progress-based geography curriculum can be planned and put into practice by primary schools. That planning for progression is problematic is recognized (Owens 2019), given the range of aspects of geography and the possible content to select from. Indeed, geography is not claimed to be an evidently progression-based subject, such as is at times proposed about mathematics. Where attempts have been made to develop a sense of progression in the subject, such as with spatial and map skills and understanding from a young age, it has not proved straightforward to clarify an agreed progression structure for teaching (Solem et al. 2014; Muniz Solari et al. 2017). It would seem that much further research is required because very little is really known about how sequences might be best described in many aspects of geography and there has been negligible investigation with younger children about how their geographical learning develops. This creates difficulties for assessing children's progressive learning in geography.

Inevitably, what has become focused on is working to state standards for children's geographical curriculum and teaching across age phases, such as in some nations' national curricula or in a national standards structure (Heffron and Downs 2012; GA 2014). The argument is that this approach supports schools by helping subject leaders and teachers to consider medium- and long-term planning in geography (GA 2014), as in global learning (Owens and Hopkin 2015). Some guidance provided for geography planning and teaching with primary children sets out a focus and structure year by year (Maude 2014), while other guidance emphasizes standards expected across two years (Catling and Willy 2018; Grigg and Hughes 2019). Rarely does it appear topic to topic, when it may not identify what is the nature of its subject progression or sequence (Scoffham and Owens 2017). A broad-based sequence approach can be helpful for schools and teachers because it specifies and clarifies benchmarks for age phases, such as by ages 7, 9, and 11 years, which may help provide a sense of progression. Figure 11.1 illustrates an aspect of the Geographical Association's 'framework for progression and assessment' as one example of identifying such benchmark standards.

The limitation in setting such expectations and benchmarks in primary geography is not that there is lack of interest in doing this. Rather, it is because there is very limited research on which to base setting such developmental standards (Muniz Solari et al. 2017). This creates a challenge for primary schools and teachers, since there are limits to what can be set out for teachers to work to within and across topics. This leads to the challenge about what it is that primary teachers actually assess about children's development and progression in geography. The reality is that this is left to chance in classrooms, unless the geography subject leader provides such guidance to staff. Yet this may depend on their knowledge of geography and how they approach progression in the subject; it may not necessarily, even then, be well-grounded since many primary subject leaders are not specialists in geography. What this identifies is that this is a ripe area for research in primary geography about what does happen in schools. While there is some doubt that setting standards, benchmarks, or expectations, year by year or across age ranges, for primary geography can be provided, it should at least be possible to identify what it is that subject leaders, teachers, and schools attempt to do to address this challenge, and how their senses of progression in geography enable them to assess children's learning developmentally or otherwise.

Seeking progress-based frameworks for primary geography teaching is important to both structuring primary geography curricula and to being able to assess children's learning within these. The purpose and value in setting out frameworks and benchmark expectations for children's geographical learning is that primary teachers can be clearer about and more focused on what they intend their children to achieve and their journey to achieve this. It will also enable primary teachers to work more effectively in providing formative and summative assessment in geography topics, assessments which support children's geographical learning. Yet, as indicated, there is much work to be done here if the challenges involved are to be overcome.

<table>
<tr><td colspan="3">Understanding of the conditions, processes, and interactions that explain features, distribution patterns, and change over time.
• Extending from the familiar and concrete to the unfamiliar and abstract;
• Making greater sense of the world by organizing and connecting information and ideas about people, places, processes, and environments;
• Working with more complex information about the world, including the relevance of people's attitudes, values, and beliefs.</td></tr>
<tr><td colspan="3">Expectations</td></tr>
<tr><td>By age 7</td><td>By age 9</td><td>By age 11</td></tr>
<tr><td>Show understanding by describing the places and features they study using simple geographical vocabulary, identifying some similarities and differences and simple patterns in the environment.</td><td>Demonstrate their knowledge and understanding of the wider world by investigating places beyond their surroundings, including human and physical features and patterns, how places change, and some links between people and environments. They become more adept at comparing places, and understand some reasons for similarities and differences.</td><td>Understand in some detail what a number of places are like, how and why they are similar and different, and how and why they are changing. They know about some spatial patterns in physical and human geography, the conditions which influence those patterns, and the processes which lead to change. They show some understanding of the links between people, places, and environments.</td></tr>
</table>

Fig. 11.1 A framework for developing the aspect of geographical understanding (GA 2014, p. 3)

Suggested research questions to pursue
Progression and assessment

In planning their school's geography curriculum, how do subject leaders identify progression? How do they see this as useful in enabling the assessment of children's geographical learning as they move through their primary years?

How do primary teachers plan for progression in their geography topics? What do they take into account in doing so, and why do they make these decisions? Are they satisfied with them? What would help them to improve their planning for progression?

Why, how, and when do primary teachers use formative, or summative, assessment to identify and enable children's progression during geography topics and across topics during a year?

> To what extent are teachers aware of progression in geographical learning throughout their school and how their assessment of their children's learning enables their children to make progress to and in the next class or stage in schooling? How do teachers build on what they are told about their new children's prior achievements and needs in geography early in the year?
>
> How, and in which aspects of geography, is the geography curriculum planned for progression within and across years? How widely is a whole school approach realized and appreciated by staff? How does this support the teaching and assessment of children's geographical learning?

11.3.3 Undertaking Assessment

Some background to assessment approaches in primary geography has been presented and research questions raised. However, the focus of assessment for teachers is almost certainly pragmatic and about what they look for in their continuous formative assessment practice alongside the activities and tasks they provide for their children in their curriculum. This indicates that the curriculum, teaching, and assessment are intricately interlinked and that implicitly in their geography teaching primary teachers constantly monitor their children's learning, as would be expected. Guidance and advice about assessment for learning in primary geography emphasizes that primary teachers should use a variety of approaches to maintain their awareness and understanding of how their children are undertaking their geographical studies. A wide range of approaches to doing this are advocated, some of which are noted in Fig. 11.2. While some of these approaches involve the teacher standing back from direct involvement with children, for instance observing how they are tackling a task or listening in to a particular discussion, others require teachers to interact with the children, such as in questioning them about their knowledge, interpretations, and feelings about a particular aspect of their geographical tasks and learning. The latter approaches may lead to setting or resetting a task, intervening to provide some direct teaching or other input that is evidently needed, or to challenging expressed information or attitudes and offering other perspectives.

While many of the assessment approaches listed in Fig. 11.2 are used as part of teachers' assessment for learning, some can be applied in summative assessment during or at the end of a geography topic. These tend to focus on whether children can recall information accurately, for instance about the locations of features and places nationally or globally, are able in a new context or with a new resource to apply their learning, such as in identifying geographical patterns on a different map, or write a paragraph explaining why they have a particular sense of or view about a place. Children may write such assessments by hand or respond to tasks set digitally by their teacher. Primary teachers might also use, possibly allied to a commercially produced and resourced geography curriculum and its topics, structured assessments

Primary teachers might draw on some or many of the following ways to monitor geographical learning:

- Observing and listening to children undertake a geography task
- Questioning children about how they are working on a particular activity and what they are learning in doing it
- Discussing with children their current activity, encouraging them to elucidate their understanding and thinking
- Explaining, and then inviting, children to give a summary of what has been explained
- Providing a resource, such as a photograph or map, and asking children what they can identify, describe, or explain about what they see in it
- As a group, asking children to discuss an aspect of the geographical topic they are learning about, encouraging each child to contribute and compare their focus of learning with each other, then sum up what they have learned
- Inviting children to reflect on a geography task they have completed, identify what they might want to learn about next, and say how this will develop their geographical knowledge
- Asking children to exchange a piece of work they are doing or have completed and to provide feedback to each other orally and/or in writing to improve or develop it
- Requiring a team of children working on a shared geography task to present it to others, such as in a debate, a drama, or an illustrated presentation, to show what they think has been most important
- Asking one child to summarize what another child has written or presented, perhaps encouraging a focus on, for instance, three points they consider to be worth knowing
- Inviting children to set their next goal(s) in their geographical learning
- Giving children a problem or conundrum to which they provide one or more responses, with reasons
- Encouraging children to talk about a place or aspect of physical or human geography they are studying to outline what they are focusing on and why
- Explaining what a particular geographical pattern or process is and why it is helpful to know about it
- Describing what different people might think about a problem or an issue and say, and even justify or challenge, why they may hold such views
- Creating a (board) game, a dilemma, or other activity to help other children understand about a geographical process, event, place, or alternative viewpoints
- Drafting and adding to a concept map to show what they know at given points in a topic from its inception to its conclusion
- Developing a campaign to alert people's attention (in school) to a particular concern and create the resources and events to maintain it over a period of time
- Maintaining a self-assessment check sheet of children's geographical learning specific to the topic, possibly across a year or longer record, of their learning confidence and needs, and being able to explain why they have made their self-assessment decisions at that point
- Identifying, justifying, and planning the next steps in an investigation or study in a geography topic

Fig. 11.2 Formative assessment approaches advised in guidance for primary geography

which are recorded and stored digitally to be retrieved. These latter assessment materials tend, it seems, to be information and skill oriented in primary geography, but this is an aspect of primary geography assessment which requires investigation to understand what teachers use and its appropriateness and effectiveness. This links, also, to what children retain as their 'residual' curriculum from their learning, what they recall sometime later. Summative assessment may or may not be informative about children's long-term, rather than short-term, geographical understanding but little is known about this among primary age children.

The problem besetting decisions about which approaches are most helpful to use and in what circumstances, and how effective they are in giving access and insight into children's geographical learning, is that there is a lack of research about how primary teachers undertake such assessment in their classes. To take one example, it has been argued that it is of value to share learning intentions with children at the start of a geography topic and for a lesson or set of lessons. What is not clear is to what extent this is done, how clear to the children the learning intentions are, and how and whether the teacher, and the children, have an evident sense that they have met these intentions during the lesson, set of lessons, or topic. This situation applies across the range and variety of approaches to formative and summative assessment advised on and no doubt used in primary geography. It is not clear in what ways providing children with oral or written feedback about their work helps them to improve it, nor whether children gain from self and peer assessment in the subject. How primary teachers undertake assessment in geography lessons is an area in need of urgent investigation.

A further and significant aspect of primary geography assessment also needs to be considered because it is essential to effective classroom assessment practices. It concerns teachers' knowledge of the subject and the language and vocabulary they use themselves and with their children. This raises the question about how a teacher's knowledge, appreciation, and valuing of geography affects the quality of their assessment practices. A teacher may state what the geography learning intentions are for a particular lesson or topic, but it is important that they understand well what they are teaching to identify and respond as children grasp ideas, make effective use of information, are able to see or not see patterns on a map, or have misunderstood what they are describing or explaining. A primary teacher needs not only to know the geography they teach but to have the vocabulary and language to recognize and develop children's good or limited and appropriate use of it, to be able to intervene to move children's learning forward, and address difficulties they may exhibit. Effective assessment, in geography as in other subjects, requires strong teacher subject knowledge, an area about which there are evident concerns (Ofsted 2011) and which is an under-researched area, most especially its impact in geography teaching and assessment.

Primary teachers' knowledge and understanding of geography lies at the heart of providing feedback to children about their geographical learning. The essence of good assessment feedback is that it recognizes the point which the child's learning has reached and gives them secure appreciation of this, as well as identifies with or for them their next steps to take. Feedback can take a variety of forms in geography lessons. As has been indicated, it might be through oral comments to or in conversation with a child or group, even to a whole class, or it may be through written responses on particular pieces of work or about a group activity such as a presentation. It needs to reflect on content and understanding to be helpful to children. Children value feedback where it is affirmative about what they can do and where it provides clarity for them about what they have found challenging and difficult, and it needs to be followed up. Evidently feedback must be based in good geographical knowledge by the teacher for both the topic being studied and the most helpful ways in which feedback can be expressed, for example in the vocabulary a teacher uses or the way in which next steps are structured for a particular child. Feedback, therefore, requires teachers to know about the strengths and needs of each child in geography. The term 'feedback' implies that it is a reflection on what the child has done. The term 'feed forward' is focused on considering the next steps in geographical learning within topics and from topic to topic. Here feed forward concerns setting targets and dialogue with children about ways they can help set their own goals. As with much else about feedback and feed forward, next to nothing is known about how this works in practice in primary geography; it is another focus which requires research.

Suggested research questions to pursue
Approaches to assessment in class

How do primary teachers approach assessment during their geography lessons and topics? What variety of means of assessing children's continuing learning do they use? To what extent do their formative approaches to assessing in geography reflect or differ from how they assess in other subject lessons and content?

What are teachers considering or looking for when they engage in formative assessment, such as through listening, by intervening, engaging in discussion, by asking questions, or by setting tasks? How are such assessments intended and planned for, and to what extent and why may they be on-the-spot decisions? What effect do they have on the children's geographical learning?

Which ways used to assess in primary geography have the most helpful impacts for children? Are some assessment methods and activities more effective than others, and does this vary between primary teachers? Why might this be? Are there significant differences between the uses and impacts of formative and summative assessments and what are these?

How does primary teachers' knowledge and confidence affect their capability to assess children's geographical learning, achievements, and needs? How does their knowledge of geography affect their decisions and choices in providing assessment activities? What are teachers concerns about the impact of their subject knowledge and confidence on their assessment practices?

In what ways do teachers provide feedback to their children about their geographical learning? How is this provided so that it is understood and appreciated by children? In what ways does it provide for next steps and target setting, acting as feed forward? What are children's responses to feedback and feed forward, and does it enable their future focus and learning in geography?

In what ways are geography assessments provided digitally, what are their geographical foci, how are they used by teachers, how do primary children respond to them, and how informative are they about children's learning and for studies they move on to?

11.3.4 Records and Accountability

Primary teachers maintain some records of children's work. These may be in portfolios with some written notes and dates, or they may be written or digital records. Some geography guidance for teachers argues that keeping cumulative and periodic records is important, while others advise that records should only be made termly or annually. Record keeping is a demanding, though necessary, task for primary teachers who may need to maintain records across the full curriculum they teach but in which there are particular demands to provide consistent and detailed records in subjects such as the national language and mathematics. Assessment records may contain remarks about particular aspects of geographical learning, even particular points children have made, may indicate on a chart children's emerging, developing, and achieving understanding of key aspects of geographical knowledge or skills, note in some way a child's attitude to geography and their task-oriented approach, or refer only to the range of topics in the subject they have covered with no indication of engagement, understanding, achievement, or needs. In some schools, children may be involved in recording their own achievements, needs, progress, and engagement in geography, again perhaps using a check-sheet and possibly adding reflective comments about what they have enjoyed or not. They may be responsible for keeping this record on paper or digitally. One approach, for example, connects gathering children's perspectives on what they say they already *know* with their sense of what they *want* to learn about, *how* they will do so, and to reviewing what they have *learnt* at the end of a geography topic—the *KWHL* approach (Grigg and Hughes 2019; Greenwood 2019)—enabling children to develop a record which allows them to compare their gains with their intentions and consider what they intended to do and did to achieve them.

While there is advice about what might be recorded, there is a lack of research into record keeping in primary geography teaching and of children's learning and progress in and across their primary years. It is, therefore, not known what may be accessible and useful forms of record keeping, by both teachers and children. The value of record keeping is that it should inform teachers and, indeed, children about what has been learned, as the basis for the children moving forward in class, from one class to the next, and for transition to their next school, though geography records have not always been of interest to secondary schools. Children's geography records may be used to provide a progress or summative report for parents and for the geography subject leader and headteacher, as a way of providing information about what children have gained or not from their geography lessons. Records can be used for teacher subject teaching accountability and staff development needs within and across subjects. The move to keeping digital records of children's geographical learning can provide greater accessibility to the subject leader as well as for passing records from teacher to teacher. But these practices and their benefits and limitations are unresearched aspects of assessment in primary geography. As in so much, guidance and advice rely only on expertise, which may be helpful but which is not supported by thorough cross-school investigations and findings.

A further approach used in primary schools to demonstrate the state and quality of geographical studies is in the display of children's work. This can be an important means of motivating children as well as a way to inform other children and teachers about work undertaken in a class. Equally, it can be informative for parents and other visitors, in this sense acting as a form of accountability through showing a range—or perhaps only the best—of work undertaken in geography by children. In this sense it can be seen as an outward-looking aspect of geography assessment and recording, which shows that, alongside other subjects similarly on show, geography has an evident and effective place in class and the school's curriculum, and that the children's work is valued. Yet, the extent to which this is undertaken, how effective it is and for whom, and what it really shows about children's geographical learning remains unclear because it remains uninvestigated.

The evaluation of the quality of geography across a school depends in part on the assessment of children's learning being effectively undertaken in all classes. Assessment forms one element among considering the geography curriculum, observations of its teaching, its available resources, the visibility of geography up and around the school, and children's perspectives on their geography education. The children's achievements and needs which geography assessments and records demonstrate provide one of the insights which senior management teams and external inspectors of schools use when making judgements about the quality of geography in a primary school and to identify areas of achievement and for improvement and future development. Maintaining an effective assessment process and record keeping in geography through a school is significant in its accountability processes but is yet to be researched.

Suggested research questions to pursue
Recording and accountability in primary geography

What is the focus of and what forms of record of children's geographical work are kept by primary teachers? How frequently are records updated and what are they used for by teachers?

To what extent are children involved in maintaining records of their geographical learning, achievements, and needs? How do they or their teachers use these? What do such records reflect about children's knowledge, skills, values, and attitudes about geography? How much do children value them and use them?

In what ways do geography records match with or link to the school and class geography curriculum? Are some types, format, and content of geography records more helpful for teachers and children than others and, if so, in what ways?

Are approaches to geography recordkeeping, such as digitally in a school system, applied across a school and accessible to teachers as well as to the subject leader and the senior leadership team? How informative are they and in what ways are they used?

How visible is children's geographical work in classes and around the school? Why is this work on display, and how was it selected? What impact does it have and on whom?

In terms of information and accountability, what types of geography records do subject leaders and senior management teams prefer access to and why? What use is made of such records?

In what ways does geography assessment contribute to the evaluation and improvement of geographical provision, teaching, and learning and their quality across a school? To what extent is assessment and record keeping a helpful element in and for geography's accountability in a school, and why is this?

11.3.5 Geography Assessment in Primary Teacher Education

There is a final vexed question focused on primary teachers' assessment in geography. It concerns primary teachers' need to learn to appreciate and undertake geography assessment during their pre-service teacher education and that they continue to develop their skills in doing so throughout their careers. Future teachers' preparation to teach tends to take place in or linked with universities and teachers' colleges. Their career professional development may be provided by these institutions, by school education authorities, by schools or groups of schools themselves, or by commercial providers. It is in these contexts that primary teachers' knowledge and skills in assessing younger children's geography should be developed and enhanced. Unfortunately, as with much else about primary teachers' knowledge of geography assessment, there is a lack of research about these aspects of their teacher preparation and continuing professional development.

When asked about the contents of their preparatory programmes for England's primary teachers, course tutors did not refer to developing pre-service teachers' skills in geography assessment in their courses (Catling 2017). It may be that assessment is a component in some geography courses, and it might be that during their pre-service future programmes teachers may be referred to guidance on teaching geography, humanities, and social studies which provide perspectives on and approaches to assessment (e.g., Barlow and Whitehouse 2019; Buchanan 2013; Brophy et al. 2013; Reynolds 2019). Texts for primary teachers generally include chapters on the assessment of children's learning in geography, humanities, and of social studies, though in the latter they may not refer to geography directly, and these chapters, or sub-sections of them, may be brief or negligible (Boyle-Baise and Zevin 2014; Grant 2017; Scoffham and Owens 2017). In other words, when future teachers search for guidance, what they encounter is limited, even on the websites of geography, humanities, and social subjects' organizations, such as the UK's Geographical Association or the Australian Geography Teachers' Association. This reinforces the concern about the dearth of studies to aid teachers in their search for research and evidence-based guidance.

This situation needs to be recognized as a key concern in university and teachers' college preparation of and support for primary teachers to teach geography. There are significant constraints in pre-service courses for all primary teachers (Catling 2017). It may well be that time for course elements on geography is so limited that assessment is barely or not covered because the focus is to develop prospective teachers' geographical understanding, which may be weak (Catling 2014), and needs to be bolstered in relation to primary curriculum requirements. If assessment is considered, it is likely to be brief and with limited examples of ways to assess primary children's geography, but this is unresearched and it is unclear what is covered when geography is included in the curriculum for future primary teachers. It may be the case that geography is subsumed with other subjects in more generalized teaching about humanities or social studies assessment and is not a specific focus. What may more usually occur is that future teachers will be expected to draw on advice, guidance, and approaches they encounter from other course components about assessing primary children's learning (e.g., Briggs 2019; Chiles 2020), though examples of assessment in geography are unlikely to be discussed, or to adapt ways to assess from other subjects, particularly from the central primary curriculum subjects of mathematics, the national language, and science. But this is an area in need of research.

For early career and experienced primary teachers, it may be the case that they have opportunities during their careers to take short courses on the assessment of primary children's geographical learning but these are likely to be taken up by only a small number of primary teachers, given the range of subject courses primary teachers could take. Such courses can be found, where they exist, through subject associations' and other providers' websites, and supportive guidance may be found in books on teaching primary geography (Catling and Willy 2018; Owens 2019). In school, it can be that the teacher who leads geography, humanities, or social studies will support their colleagues and may provide sessions about assessing geographical learning. Again, this is an unresearched area. The key message is that next to nothing is known about the extent to and ways in which teachers are prepared for and enabled to enhance their capabilities to undertake specific geography assessment with primary children of all ages.

Suggested research questions to pursue

Geography assessment in primary pre-service and professional development programmes

In pre-service courses for primary teachers, are students taught about the assessment of children's geographical learning, whether this is in a geography, humanities, or social studies? Or are they expected to transfer their learning in other subjects or from generalized contexts into their geography teaching? When in their programme might pre-service teachers be taught about primary geography assessment? What expectations are there about their learning about geography assessment?

Which aspects of geography assessment are covered? How are pre-service teachers taught about geography assessment and what variety of examples are used by tutors? Are these drawn from and related to school practices? What time and resources are provided for this element of the course? Does assessing primary children's geography feature in any assessment of such a course and, if so, in what form? How do students engage with and respond to being taught about how to assess primary children's geographical learning?

What opportunities do career primary teachers have to develop their skills in primary geography assessment? How often and when do these courses occur? How accessible are they to primary teachers, or are they provided essentially for geography, humanities, and social studies leaders in primary schools? What aspects of primary geography assessment are included in such courses, and how do they relate to the needs and demands of primary schools, for instance in relation to geography curriculum requirements?

What approaches to staff development about primary geography assessment do geography, humanities, or social studies lead teachers take? When and how are such updating and development works undertaken? How do primary teachers respond to these sessions? What impact are they intended to have, and do they have, on the effectiveness of teachers' assessment of children's geographical learning?

In each of the above, what benefits and limitations can be identified? What variety of practice is there in universities and education colleges and in schools? To what extent are such practices undertaken well or evaluated as a concern in pre-service teacher education and continuing professional development? How might these matters be enhanced or addressed?

11.4 Conclusion: Future Direction for Primary Geography Assessment

As is now evident there is much to consider about assessment in primary geography. While there is a great deal of advice about what to focus on, the opportunities for assessment, the variety of assessment practices and tasks, and how these might be of value for children and informative for teachers and schools, there remains a lack of research about assessment in primary geography. This is not to say that effective and good assessment purposes and practices are not used in many primary schools and classrooms, but the problem lies in not knowing what these are nor what best practices might be. Because a school or teacher may feel comfortable undertaking assessment in their geography teaching in the ways that they do does not indicate that what is done is as informative as it needs to be for the teacher to progress learning nor in the best interests of and for the learning needs of the children.

In examining what is needed in geography education research, the *Committee of the Road Map for the 21st Century Geography Education Project* in the USA identified assessment in geography as a key area for significant investigation and recommendations (Bednarz et al. 2013). The separate assessment report focused to a large extent on matters of the formal design and the quality of large-scale assessments, some provision for which there has been towards the end of elementary (primary) education (Edelson et al. 2013). Assessment for learning (formative assessment) is given little space in the assessment report, though its importance in classroom teaching is stressed through its relationship to the goals of geography lessons, its need to be fine-grained to be of value for checking that learning has taken place or not, and to be well-timed. The report noted that assessments can be a distraction from teaching, while other research in the wider context of social studies teaching (of which geography forms an element in the USA and elsewhere) states that there can often be difficulties 'untangling' formative assessments from primary teachers' classroom practices (Grant 2017). Nonetheless, it is recognized that in reality little is known about teachers' assessment for learning in primary geography or social studies. Indeed, a reading of both studies indicates an emphasis on the assessment of geographical information, such as place locations or physical geography processes in standards assessments, rather than on the deeper aspects of thematic and conceptual understanding with younger children. The accessible, externally-to-schools produced formal, national assessments were identified as unreliable and of poor validity because of concerns about the imbalance in the aspects of geography assessed, containing poorly written and largely recall-only questions, and being inappropriate for the geography curriculum which supposedly they assessed. However, alongside this, Grant (2017) raised questions about the reliability and validity in classroom teachers' assessment of children's (geography within) social studies learning, which has been the focus of this chapter. Such matters as these, and the nature of the geographical knowledge taught and assessed in the primary classroom, require further examination and investigation.

While there is probably much that is helpful and effective in primary classroom teachers' assessment of their children's geographical learning, little really is known because of the lack of research about geography assessment in the primary classroom and school context. It is one thing to think that assessment may be done well; it is another to evidence that it is. In England, assessment in primary geography has been an area of concern identified through the school inspection process (Ofsted 2011). It is stated to be a weakness in many primary teachers' practice, which needs support and guidance. This chapter has drawn on the range of guidance about assessment which specialists in primary geography and areas of social studies and humanities have provided in several countries, though this is meagre. Indeed, not all guidance about teaching and learning geography discusses assessment (Pike 2016; Dolan 2020). The foregoing discussion has identified that much of the advice is not underpinned by research in and about primary geography assessment in classrooms. This has indicated a variety of areas as an agenda for such research in various aspects of primary geography assessment:

- The purposes of assessing geography in primary children's learning
- Diagnostic assessment for and in primary geography
- The interplay between and quality of primary geography planning and assessment
- The relationship of progression and assessment in primary geography
- Teachers' approaches to and means of geography assessment in the primary classroom

- The nature and uses of primary geography record keeping and displays
- The role and value of assessment in primary geography accountability and evaluation.

It has also indicated that this agenda needs to include:

- The provision in and nature of the pre-service teacher education and the continuing professional development of teachers' knowledge about and practices in assessing primary children's geographical learning.

This is by no means a complete list. To it might be added:

- Whether primary teachers make use of guidance about geography curriculum, teaching, and assessment
- The impact of teachers' subject knowledge, confidence, and interest on the quality of their assessment of children's geographical learning
- What the assessment needs and challenges are in relation to the range of geography's subject content for primary children, including assessments of geographic information, thematic aspects, and conceptual thinking
- How primary geography provides insight for teachers into their children's understanding of geography beyond the factual information that the subject utilizes about places and human and physical geography
- How children apply their geographical learning across the geography topics they study during their primary school career
- Aspects of children's lives and learning which assist or inhibit them in their geography learning and assessment
- Whether assessment concerns about primary geography are specific to geography or are general across subjects, why this is, what needs to be done to address this and how it can be undertaken.

No doubt there is more!

While this chapter has indicated the range of elements involved in primary geography assessment, it has provided an extensive agenda for research about primary geography assessment. This needs careful consideration and priorities need to be set. Given that so little is known, the four core needs to investigate across primary teachers and schools would appear to be as follows. One is to investigate what primary teachers intend assessment in their geography planning and teaching will enable them to achieve for their children. The second is to delve into the formative and summative practices teachers use to help and develop children's in-class learning in geography, informally and formally. The third is how their assessment intentions and practices relate effectively to the geography curriculum topics they teach. The fourth is how what they do and learn about their children's geography enables them to improve the quality of their geography teaching and children's learning. In essence such research will focus on the two key foci of primary geography assessment. The first is about the value of the geography curriculum which teachers, schools, and external bodies provide and set out for children during their primary schooling, even when it includes opportunities for children's contributions; that is, the worthwhileness of what is taught. The second is whether and how children benefit from their geography curriculum, not least through consideration about whether they have gained insights into how the world works locally and globally, and what it means for and how it impacts on them and their communities. The value of assessment research must be to develop ever more effectively children's learning in and of geography and teachers' capability to provide for this.

References

Alexander R (ed) (2010) Children, their world, their education. Routledge, Abingdon

Assessment Reform Group [ARG] (2005) The role of teachers in the assessment of learning. London University Institute of Education, London

Barlow A, Whitehouse S (2019) Mastering primary geography. Bloomsbury, London

Bednarz S, Heffron S, Tu Huynh N (eds) (2013) A road map for 21st century geography education: geography education research. National Geographic Society, Washington http://natgeoed.org/roadmap

Blundell D (2016) Rethinking children's spaces and places. Bloomsbury, London

Boyle-Baise M, Zevin J (2014) Young citizens of the world: teaching elementary social studies through civic engagement, 2nd edn. Routledge, New York

Briggs M (2019) What is assessment? In: Carden C (ed) Primary teaching. Sage, London, pp 279–298

Brophy J, Alleman J, Halvorsen A-L (2013) Powerful social studies for elementary students, 3rd edn. Wadsworth, Belmont

Buchanan J (2013) History, geography and civics: teaching and learning in the primary years. Cambridge University Press, Port Melbourne

Butt G (2020) Geography education research in the UK: retrospect and prospect—the UK case within the global context. Springer, Cham

Catling S (2013) Editorial: the need to develop research in primary children's and schools' geography. Int Res Geogr Environ Educ 22(3):177–182

Catling S (2014) Pre-service primary teachers' knowledge and understanding of geography and its teaching: a review. Rev Int Geogr Educ Online 4(3):235–260

Catling S (2017) Not nearly enough geography! University provision for England's pre-service primary teachers. J Geogr High Educ 41(3):434–458

Catling S, Willy T (2018) Understanding and teaching primary geography. Sage, London

Chawla L (ed) (2002) Growing up in an urbanising world. UNESCO–Earthscan, London

Chiles M (2020) The CRAFT of assessment. John Catt Educational, Woodbridge

Christodoulou D (2016) Making good progress? The future of assessment for learning. Oxford University Press, Oxford

Clarke S (2014) Outstanding formative assessment: culture and practice. Hodder Education, London

Dann R (2014) Assessment as learning: blurring the boundaries of assessment and learning for theory, policy and practice. Assess Educ Princ Pol Pract 21(2):149–166

Dolan A (2020) Powerful primary geography: a toolkit for 21st-century learning. Routledge, Abingdon

Donarski S (2020) The researchED guide to assessment: an evidence-informed guide for teachers. John Catt Educational, Woodbridge

Edelson D, Shavelson R, Wertheim J (2013) A road map for 21st century geography education: assessment. National Geographic Society, Washington http://natgeoed.org/roadmap

Faragher S (2014) Understanding assessment in primary education. Sage, London

Freeman C, Tranter P (2011) Children and their urban environment: changing worlds. Earthscan, London

Freeman C, van Heezik Y (2018) Children, nature and cities: rethinking the connections. Routledge, Abingdon

Gardner J (ed) (2012) Assessment and learning, 2nd edn. Sage, London

Geographical Association [GA] (2014) An assessment and progression framework for geography. www.geography.org.uk/curriculum2014/assessment

Geographical Association [GA] (2015) Assessment—practical steps to support progression and attainment in geography. www.geography.org.uk/Assessment-without-levels-practical-steps-to-support-progression-and-attainment-in-geography

Grant S (2017) The problem of knowing what students know: classroom-based and large-scale assessment in social studies. In: Manfra M, Bolick C (eds) The Wiley handbook of social studies research. Wiley, Chichester, pp 461–476

Grant S, Salinas C (2008) Assessment and accountability in social studies. In: Levstik L, Tyson C (eds) Handbook of research in social studies education. Routledge, New York, pp 219–236

Greenwood R (2019) KWL grids: pupil voice in action. Prim Geogr 98:7–8

Griffiths J (2013) Kith: the riddle of the childscape. Hamish Hamilton, London

Grigg R, Hughes S (2019) Teaching primary humanities, 2nd edn. Routledge, Abingdon

Harwood D, Rawlings K (2001) Assessing young children's freehand sketch maps of the world. Int Res Geogr Environ Educ 10(1):20–45

Hayward L, Hayward S (2016) Assessment and learning. In: Wyse D, Rogers S (eds) A guide to early years and primary teaching. Sage, London, pp 165–183

Heffron S, Downs R (eds) (2012) Geography for life: national geography standards, 2nd edn. National Geographic Council, Washington DC

Holt L (2011) Geographies of children, youth and families: an international perspective. Routledge, Abingdon

Lane R, Bourke T (2019) Assessment in geography education: a systematic review. Int Res Geogr Environ Educ 28(1):22–36

Malone K (2018) Children in the Anthropocene: rethinking sustainability and child friendliness in cities. Palgrave Macmillan, London

Marcinkowski T, Shin D, Noh K-I, Negev M, Sagy G, Garb Y, McBeth B, Hungerford H, Volk T, Meyers R, Erdogan M (2013) National assessments of environmental literacy: a review, comparison and analysis. Routledge, New York

Matthews H (1992) Making sense of place: children's understanding of large-scale environments. Harvester Wheatsheaf, Hemel Hempstead

Maude A (2014) Understanding and teaching the Australian curriculum: geography for primary schools. Hawker Brownlow, Melbourne

Muniz Solari O, Solem M, Boehm R (eds) (2017) Learning progressions in geography education: international perspectives. Springer, Cham

Nairn K, Kraftl P (eds) (2016) Space, place and environment. Springer, Singapore

Ofsted (2011) Geography: learning to make a world of difference. www.ofsted.gov.uk/publications/geography-learning-to-make-a-world-of-difference

Owens P (2019) Effective subject leadership. In: Willy T (ed) Leading primary geography. Geographical Association, Sheffield, pp 112–157

Owens P, Hopkin J (2015) Progression in Global Learning. Prim Geogr 87:6–7

Peacock A (2016) Assessment for learning without limits. Open University Press, London

Pickford T, Garner W, Jackson E (2013) Primary humanities: learning through enquiry. Sage, London

Pike S (2016) Learning primary geography. Routledge, Abingdon

Pollard A (ed) (2019) Reflective teaching in schools. Bloomsbury, London

Reynolds R (2019) Teaching humanities and social sciences in the primary school, 4th edn. Oxford University Press, Melbourne

Scoffham S (2019) The world in their heads: children's ideas about other nations, peoples and cultures. Int Res Geogr Environ Educ 28(2):89–102
Scoffham S, Owens P (2017) Teaching primary geography. Bloomsbury, London
Solem M, Tu Huynh N, Boehm R (eds) (2014) GeoProgressions: learning progressions for maps, geospatial technology and spatial thinking: a research handbook. Association of American Geographers, Washington, DC
Spendlove D (2009) Putting assessment for learning into practice. Bloomsbury, London
Wiliam D (2011) What is assessment for learning? Stud Educ Eval 37(1):3–14

Simon Catling is Emeritus Professor of Primary Education, Oxford Brookes University. A Past President of the UK's Geographical Association, he has been engaged in researching and promoting primary geography nationally and internationally for many years, publishing widely. Following teaching in three primary schools, including deputy headship, he led courses for future and current teachers in primary geography throughout his career at OBU, alongside his roles in university department leadership and management. He maintains his interests and studies in children's geographies, curriculum, teaching and learning in primary geography, and geography textbooks.

GPSR Compliance

The European Union's (EU) General Product Safety Regulation (GPSR) is a set of rules that requires consumer products to be safe and our obligations to ensure this.

If you have any concerns about our products, you can contact us on ProductSafety@springernature.com

In case Publisher is established outside the EU, the EU authorized representative is:

Springer Nature Customer Service Center GmbH
Europaplatz 3
69115 Heidelberg, Germany

Batch number: 10726135

Printed by Printforce, the Netherlands